THE
NEXT
WAVE

BROOKINGS FOCUS BOOKS

Brooking Focus Books feature concise, accessible, and timely assessment of pressing policy issues of interest to a broad audience. Each book includes recommendations for action on the issue discussed.

Also in this series:

Fast Forward: Ethics and Politics in the Age of Global Warming
by William Antholis and Strobe Talbott

Brain Gain: Rethinking U.S. Immigration Policy
by Darrell M. West

The Pursuit of Happiness: Toward an Economy of Well-Being
by Carol Graham

A BROOKINGS FOCUS BOOK

THE
NEXT
WAVE

USING DIGITAL TECHNOLOGY TO FURTHER SOCIAL AND POLITICAL INNOVATION

Darrell M. West

BROOKINGS INSTITUTION PRESS
Washington, D.C.

Copyright © 2011
THE BROOKINGS INSTITUTION
1775 Massachusetts Avenue, N.W., Washington, D.C. 20036
www.brookings.edu

Library of Congress Cataloging-in-Publication data
West, Darrell M., 1954–
The next wave : using digital technology to further social and political
innovation / Darrell M. West.
    p.   cm.
Includes bibliographical references and index.
Summary: "Maintains that the digital revolution is well under way and will
continue to lead to important changes in the way people, the for-profit and
nonprofit sectors, and government will function and interact, with emphasis
on the importance of innovation and the need to balance innovation, privacy,
and security"—Provided by publisher.
ISBN 978-0-8157-2188-8 (cloth : alk. paper)
  1. Information technology—Economic aspects. 2. Technological innovations
—Economic aspects. I. Title.
HC79.I55W488 2011
  303.48'33—dc23   2011012918

9 8 7 6 5 4 3 2 1

Printed on acid-free paper

Typeset in Sabon

Composition by Cynthia Stock
Silver Spring, Maryland

Printed by R. R. Donnelley
Harrisonburg, Virginia

To the high-tech entrepreneurs

who fueled the digital revolution

CONTENTS

CONTENTS

IV. NORMATIVE CONCERNS

V. CONCLUSION

ACKNOWLEDGMENTS

I WANT TO ACKNOWLEDGE THE help of several individuals and organizations. Valuable research assistance on this project was provided by Jenny Lu and Raffaela Wakeman, to whom I am very grateful. Jenny Lu made major contributions to chapter 6 on innovation in the public and private sectors. Russ Whitehurst and E. J. Dionne wrote a policy report that went into parts of chapter 9, on new digital media. Allan Friedman co-wrote a report on cloud security that forms part of chapter 10. The Brookings Governance Studies program and its Center for Technology Innovation provided time and resources for policy reports that formed the basis for some of the book's chapters. The Ewing Marion Kauffman Foundation and the Bill and Melinda Gates Foundation provided financial support for portions of this book. I appreciate the helpful comments of reviewers Kevin Esterling and Helen Margetts on the book manuscript. Their suggestions regarding organization and argument substantially improved the final product. None of these individuals or organizations are responsible for the interpretations presented here.

THE
NEXT
WAVE

CHAPTER ONE

A FRAMEWORK FOR ANALYSIS

A HEATED DEBATE OVER NEW technology unfolded on the floor of the U.S. Senate in 1930. Some senators were incensed because the Chesapeake and Potomac Telephone Company wanted to replace operator-assisted phones with a new device called the rotary dial telephone. Rather than using a human operator to place a call, legislators would have to dial their own calls.

Senator Carter Glass complained that dial phones were overly complicated and "difficult to operate." Senator Clarence Dill also commented on the difficulty in using the new phones: "One has to use both hands to dial," he lamented. In addition, users "must be in a position where there is good light, day or night, in order to see the number; and if he happens to turn the dial not quite far enough, then he gets a wrong connection."[1] Dill claimed that the new phones were especially difficult for older members of Congress.

Other senators challenged these assertions and argued that the new phones actually were easier to use. Senator Millard Tydings noted that legislators could place calls themselves and without going through a central switchboard. It was a more efficient way to work, and it gave callers more control over their telephoning, he noted.

At the end of the debate, the majority was not persuaded. The Senate passed a resolution delaying adoption of the new communications system. The stalemate was not broken until the phone company announced a compromise. It offered to install either dial or operator phones for individual senators, according to their wishes. Thus technology innovation could proceed in a way that was compatible with the preferences of individual members.

The controversy over rotary telephones was not unusual. New technologies often inflame passions, divide people, and raise questions regarding personal and societal impact. By their very nature, they are disruptive because of their ramifications for the way society and government function and the manner in which people interact with one another. Proponents of technology always tout a range of benefits, while critics worry about negative consequences for social life, political institutions, and individual values.

This book does not seek to resolve fundamental disputes over the value of every invention. Rather, it focuses on how the next wave of digital technologies can be harnessed to further U.S. social and political innovation. My goal is to explore a range of specific developments in the contemporary period, analyze what they mean for individuals, society, and government, and understand which barriers limit their enactment.

There are many possibilities for digital technology that are consistent with personal and social values. Governments and other organizations can employ new advances that allow them to perform faster, smarter, and more efficiently. People can deploy digital technology to improve transparency, participation, and collaboration. By working with the private sector, political leaders can stimulate a flowering of innovation in a variety of policy areas.

But proponents must be aware of public fears, institutional barriers, and the real privacy and security threats posed by digital developments. Technology rarely drives change in isolation from other forces.[2] Effective implementation arises from a combination of technology, organizational shifts, and policy reforms. The task is to further innovation while also protecting basic social and individual values.

THE VIRTUES OF TECHNOLOGY INNOVATION

Technology innovation represents one of the most important keys to long-term prosperity and competitiveness. One of the reasons why the United States thrived after World War II and in following decades was its emphasis on invention. The nation brought to its shores leading scientists from European countries and supported their basic and applied research.

The results were spectacular. The United States became a world leader in science and technology. The early Russian launch of the Sputnik satellite notwithstanding, the United States became the first nation to put a man on the moon and successfully commercialized products derived from the space program such as satellite communications, global positioning systems, and wireless communications.[3]

Scientists undertook basic work in computer science, material sciences, genomics, neuroscience, and cognitive science that launched new industries and created vast wealth. Digital technology and the life sciences spawned the computer revolution, new medical treatments, and the deciphering of the human genome, among other things. Like the electric grid of the early twentieth century and the interstate highway system of the late twentieth century, the Internet became an infrastructure platform for progress in education, health care, energy efficiency, communications, and mass entertainment.

Not surprisingly, given the potential of these technologies, a number of countries have identified broadband and wireless as crucial for national development. Broadband is viewed in many places as the key driver of economic development, social connections, and civic engagement. In a study based on the experience of 120 nations between 1980 and 2006, Christine Qiang estimates that each 10 percentage point increase in broadband penetration adds 1.3 percent to a high-income country's gross domestic product and 1.21 percent to that of low- to middle-income nations.[4]

Broadband is crucial because it is a cross-cutting technology that speeds innovation in health care, education, energy, and social networking. High-speed broadband allows physicians to share digital images with colleagues in other geographic areas. Schools are able to extend distance learning to underserved populations. Smart electric grids produce greater efficiency in monitoring energy consumption and contribute to more environment-friendly policies. Video conferencing facilities save government and businesses large amounts of money in travel expenses. New digital platforms across a variety of policy domains spur usage and innovation and bring additional people, businesses, and services into the Internet revolution.[5]

The goal of these and other new technologies is to improve communications, organizational efficiency, and individual effectiveness. By democratizing information production, technology lowers the cost of information and reduces the barriers to entry within a number of fields.[6] With modest resources, and without reliance on large-scale capital and human manpower, it is possible to create new software, devise novel applications, and bring new products to market.

THE RISKS OF TECHNOLOGY INNOVATION

While digital technologies offer many benefits, they also raise important questions about social and individual values. Many worry about privacy, security, and societal impact, among other issues. Is it possible to maintain individual privacy and security in a wired world? What happens to social organizations when interactions and social delivery move from face-to-face to electronic communications? Will health information and education technology improve treatment and learning or merely represent a new way of delivering current services?

4

Every historical era has seen major conflicts over technology innovation. The printing press is thought to have been disruptive to the established religious and political order of medieval Europe. In a world where knowledge was controlled by kings and popes, Johannes Gutenberg's invention reduced the costs of information and facilitated broader production and dissemination of knowledge. Its emergence laid the communications groundwork for what became major upheavals in religion and governance.

The same was true for radio and television. Television is blamed for a reduction in the quantity and quality of political news and the superficiality and inaccuracy of civic discourse. With its emphasis on visual communication and short sound bites, television brought new types of political leaders to the forefront and helped them win elective office. Broadcast media speeded up the news cycle and, with the addition of cable channels and opinionated hosts, undermined the civility of public discussions.

People worry about how to maintain confidentiality of online services and transactions and how to avoid the deleterious features of virtual interactions in the digital world. The emergence of spam, hackers, and coordinated security intrusions challenges electronic communications. These concerns slow the adoption of technology innovation and raise costs for consumers, businesses, and government agencies.

The key challenge for policymakers is to find ways to balance competing goals such as innovation, privacy, and security. There is an inherent tension across these dimensions. The more money spent on security, for example, the higher software costs become and the more difficult it is for small innovators to create new products.

There also is a trade-off between privacy and security. It is possible to improve computer security by allowing more monitoring of suspicious activities. But opening up back channels to surveillance compromises personal privacy and makes individual confidentiality more difficult to maintain.

HOW TO INNOVATE EFFECTIVELY

Effective digital innovation is facilitated by a combination of new technology, organizational restructuring, and policy changes. One without the others does not make for successful social or political innovation. A strategy for innovation is most effective when it alters organizational routines, redirects resources, creates economies of scale, and increases efficiency and productivity among the general workforce.[7] For example, private sector companies have been successful at using technology to drive organizational change. Businesses deploy information technology to improve worker productivity, reengineer corporate practices, and reduce hierarchy in organizations. They automate repetitive tasks, cut out middle managers, and facilitate digital communication flows across levels of the organization. These changes help businesses become more efficient without compromising their ability to deliver needed services.

Governments have been less successful in replicating this formula. Flattening organizations is difficult for public agencies because it involves laying off mid-level managers. Although many agencies perform repetitive tasks, service delivery to citizens in a public setting often involves great nuance from case to case. The complexity of the requests that come before agencies makes it difficult for them to automate. People want their individual circumstances incorporated into the government decision, not simply a uniform response. Part and parcel of government administration is discretion to adjust responses based on individual circumstances.

In addition, some government agencies are unwilling to use technology to change organizational routines. Rather than seeing technology as a chance to alter work roles and organizational practices, executives use it to reinforce existing arrangements. They do not take advantage of new delivery systems to make fundamental changes within their own agencies. This limits the transformational potential of digital government.

Yet those are exactly the types of alterations required to reap the benefits of digital technology. If agencies wish to gain the benefits of technology innovation, they need to alter organizational routines and make policy changes that facilitate the efficiencies and economies of scale made possible through new approaches.[8] This is a strategy that has worked well in the private sector, and it also has the potential to be effective in the public arena.

Limits on the ability of public organizations to embrace change does not mean that effective government innovation is impossible. Rather, the analysis of innovation in public and private sector organizations demonstrates that change is possible under given sets of conditions.[9]

The analysis suggests that there are five keys to effective innovation. First, successful adopters must devote sufficient resources to the innovation process. One of the reasons why the private sector is more successful than government agencies at technology innovation is that businesses spend a higher proportion of their overall budget on information technology. According to research I have undertaken, successful companies spend 2.5 percent of their budgets on technology, higher than the average of 1.88 percent found in state government agencies.[10] Financial resources are important because they allow innovators to undertake market research, plan effective strategies, and provide resources up front and throughout the implementation phase.

Second, successful innovators focus on the customer, value market research, and take visitor feedback seriously. In interviewing leaders in various organizations, I found that they attribute effective technology innovation to market research and an understanding of what their customers want. One government official said that he would like to do market research for his agency but lacked the resources to do so. Asked how he got feedback on what visitors liked, he said that the agency monitored its complaint lines, and when dissatisfaction rose, the agency knew it had a problem. But feedback is reactive, not proactive, and by the time complaints start coming in, it is already too late. The key to customer orientation is undertaking research that anticipates what is going to be problematic down the road. This ability to see around corners is what distinguishes successful from less effective innovators.

Third, technology innovators provide incentives for management and design teams to work together. One of the key requirements in technology innovation is getting organizational incentives right. Too many agencies do not align their management structures and design teams in a way that encourages people to work together. Successful innovation requires breaking down barriers and figuring out how to get people to cooperate. Based on interviews in the public and private sectors, the biggest barrier to innovation is unwillingness to work together across organizational domains. Reaching economies of scale, saving money, and boosting productivity cannot be achieved without organization-wide cooperation.

Fourth, innovators devote time to understanding their competition and determining how to position themselves vis-à-vis market competitors. Business leaders reported that understanding the competition is vital to effective implementation. This is why companies often have an innovation advantage over the public sector. They face competition and will lose money if they do not understand their niche relative to other companies. The problem in the public sector is that agencies generally do not have competitors and public organizations often do not have sufficient incentives to learn from other agencies. This limits their ability to adapt to changing circumstances and makes it difficult for them to adopt new practices.

Finally, successful innovators tie resource allocation to customer satisfaction. Ultimately, there must be clear consequences that result from effective or ineffective technology innovation. Positive outcomes should accrue to units that innovate, understand the competition, and undertake market research. Similarly, there must be ramifications for failure. Unless organizations incorporate consumer reactions in resource decisions, staff members will not take consumer views seriously. In the public arena, agencies can get visitor feedback in a variety of ways: online surveys, comment forms, systematic surveys regarding visitor experiences, collaborative decisionmaking, and crowd-sourcing processes. These mechanisms need to be added to organizational decisionmaking so that those agencies getting positive reviews receive a bonus for excellent performance.

CASE SELECTION

In this volume I examine a variety of different types of social and political innovation. Among the developments reviewed here are changes in public and private sector innovation (for example, electronic government and customer-driven health care), platform shifts (such as cloud computing and broadband), and social innovations (such as personalized medicine, mobile communications, and digital media).

These cases represent different types of innovation and illustrate ways to balance competing objectives. In each example, I look at what the innovation involves, what barriers complicate successful implementation, and ways to overcome real or potential problems. The goal is to understand the dynamics of innovation and policy actions that encourage progress.

Several common features of innovation occur across these disparate areas. These include strong leadership, proper financing, organizational coordination, favorable policy incentives, and a supportive political environment. In each case, I examine each of these features to understand how to innovate most effectively.

I employ a variety of methodologies to study technology innovation. These include case studies of successful innovations, interviews with leading observers, opinion surveys outlining public attitudes, and analysis of technology data. I combine different types of approaches to analyze how new technologies are transforming organizations, what factors promote innovation, and how barriers to change might be overcome.

PLAN OF THE BOOK

The book is organized in five parts dealing with technology innovation, digital platform, policy considerations, normative concerns, and a conclusion that addresses ways to facilitate innovation. Part 1 emphasizes technology advances in the public and private sectors. In chapter 2, I look at the public sector and examine how technology can make government faster, smarter, and more efficient. I argue that technology changes the relationship between citizens and leaders and creates opportunities for enhanced transparency, participation, collaboration, and social networking. In this chapter, I review the manner in which technology enables agency change and furthers citizen engagement with government and explore the barriers to public sector innovation.

Chapter 3 examines private sector technology innovation in the realm of health care. A revolution is taking place in medical care, powered in part by technology that is empowering ordinary patients. Consumers have access to more information than ever before, and they are using new digital tools to share cost information and learn about treatment options. This shift breaks down the hierarchy that has dominated the industry and gives people more control over their health care. In this chapter, I look at customer-driven health care and the opportunities and limits it presents.

Part 2 focuses on platform shifts to cloud computing and broadband. Chapter 4 examines ways public agencies can save money through cloud computing. I review examples of successful migration from local to remote file servers and how government officials have saved substantial amounts of money. Through more efficient file storage, reduced personnel costs, and more effective use of technology resources, cloud computing can safeguard financial resources and produce better results for citizens and government officials. There are, however, privacy and security concerns that complicate the future of cloud computing.

Chapter 5 investigates innovation through the adoption of high-speed broadband. I examine the experience of four countries in broadband development, both new efficiencies and economies they have achieved and problems they are encountering. I review cases of effective implementation and the new applications that have been developed elsewhere.

Part 3 looks at policy aspects of innovation. Chapter 6 examines the differential rate of technology innovation between the public and private sectors. In many respects, the public sector has been slower to innovate than the business world. I compare innovation across government agencies and leading companies and show why government lags, what values it emphasizes that are sometimes ignored in the private sector, and what it can do to catch up.

Chapter 7 looks at mobile communications and consumer experiences of that technology. The rise of handheld devices has altered the way people acquire information and use digital technology. Information and services are increasingly being accessed through mobile devices. However, not all consumers are happy with their mobile experiences. In this chapter, I review survey data regarding mobile communications and discuss some of the changes users would like to see in their mobile plans and what they are willing to pay for various services.

Part 4 explores the normative aspects of innovation. Chapter 8 focuses on personalized medicine and health information technology. Advances in human genomics and information technology make it possible to envision a situation in which treatments are tailored to individual genetic structures, prescriptions are analyzed in advance for likely effectiveness, and researchers study clinical data in real time to learn what works. Yet there are privacy and confidentiality concerns that affect people's views about genomics and health information technology. In this chapter, I review normative and operational aspects of personalized medicine and what changes are required to increase the use of genomics and health information technology.

Chapter 9 investigates the future of the new media and the way people acquire information. Traditional news organizations have faced a perfect storm of financial disaster. Between new technologies and disruptions of their previous business models, they are reinventing themselves and using the Internet as their major delivery system. This has altered the way the news media function and the manner in which people get information. I explore what this new media universe looks like, what risks and opportunities are present, and what steps are needed to strengthen news organizations.

Chapter 10 examines the challenges raised by digital technology for privacy and security. New platforms generate present and potential problems in maintaining confidentiality in a digital era and protecting consumers from corporate, government, and international snooping. New technologies create opportunities for individual action but entail risks in terms of safeguarding basic rights. I examine ways to maintain privacy and security in a digital age.

The final part presents a summary of the book and addresses the transformational potential of new technologies. I discuss what must be done to encourage future innovation. The agenda for change must encourage entrepreneurship and innovation and align organizational structure with technology innovation. I outline concrete policy actions needed to allow Americans to gain the full advantages of digital technologies.

PART ONE

TECHNOLOGY INNOVATION

CHAPTER TWO

IMPROVING GOVERNMENT PERFORMANCE

AMERICAN PUBLIC SECTOR PERFORMANCE LEAVES much to be desired. A 2010 CNN/Opinion Research survey, for example, has found that 86 percent of Americans believe the federal government is broken.[1] Trust in government is at an all-time low. In one 1964 survey, 76 percent of respondents trusted government just about always or most of the time. Now, that number is down to 18 percent.[2] The public opinion climate has shifted from one in which most Americans trust government to do the right thing to one where mistrust is rampant.[3]

There are many reasons for the public dissatisfaction that has arisen over the past half century: personal scandals, official misconduct, high unemployment, unpopular wars, large deficits, inadequate schools, and the soaring costs of health care, among others. Many citizens do not see our elected officials doing much to solve problems. The United States has weak leadership, strong political polarization, superficial media coverage, and limited public understanding of difficult policy issues. This constellation of personal and institutional failures makes it difficult to address the long-term problems that face the United States.

Neither party has a monopoly on poor government performance. In the past few years, governing challenges have complicated our ability to address a variety of issues, such as the economy, health care, financial regulation, climate change, and immigration. But these governance difficulties are not unique to the Obama administration. Most recently, examples of governability challenges were common during the presidencies of George W. Bush (social security, tax reform, and immigration) and William J. Clinton (health care, the environment, and trade-negotiating authority).

Given the numerous examples of dysfunctional politics leading to broken policies, from both parties and over an extended period, we need to reform our political system in ways that enhance our capacity for leadership, innovation, and problem solving. We must consider new approaches to public sector innovation that make government faster, smarter, and more efficient.

IMPROVING EFFICIENCY AND PRODUCTIVITY

Some think it is impossible to reform government, that the public sector is and will always be inefficient and wasteful. Critics argue that it is difficult to introduce accountability, responsiveness, and productivity into the public sector for several reasons. Government agencies are large, bureaucratic organizations and therefore difficult to change. They have few market incentives to become more accountable, responsive, and productive. Finally, government offices have monopoly power in particular areas and do not face the competition that would require them to improve efficiency and effectiveness.

Yet large, private organizations have seen dramatic improvements in productivity and efficiency, even companies with cumbersome bureaucratic operations. Through a combination of new technology and new organizational routines, many businesses have fundamentally altered the way information and services are accessed. In banking, for example, customers who once used human tellers to make deposits and withdrawals now turn to automated teller machines. In some stores, self-service scanners combined with debit or credit card readers mean consumers can make purchases with no human interaction at all. At airports, there has been a shift from manual to computerized check-in procedures.

According to one report, 70 percent of air travelers now use automated kiosks to check in for flights. Two-thirds say they are willing to use automation for airport tasks such as "paying baggage fees, buying meals, printing bag tags, and claiming delayed baggage." Nearly one-quarter indicate they have checked in via mobile phone.[4]

These types of technologies save money, reduce personnel needs for routine tasks, and improve organizational efficiency and productivity. There are many success stories from the private sector of better performance through digital technology and organizational change.[5] The question is how best to facilitate innovation in the public sector.

Old Approach to Government Reform: Move Boxes Around or Create New Boxes

Historically, there have been two approaches to U.S. government reform.[6] The first is to reorganize existing agencies. In 1979, for example, the Department of Health, Education, and Welfare was transformed into Health and Human Services. There have been reorganizations of the Central Intelligence Agency, the Department of Defense, and the Federal Emergency Management Agency. Reorganizations often are popular because people equate movement with action. But moving government boxes around does not fix problems.

The other approach taken is to create new departments or agencies. In 1970 Congress established the Environmental Protection Agency to deal with an area perceived to be critically important but neglected by existing public organizations. Following the September 11, 2001, terrorist attacks, Congress mandated a new Department of Homeland Security to supervise national protection. Public officials hope that new structures will attract better financial resources, greater public support, stronger political will, and extra authority to address newly emerging problems.

New Approach: Transforming Operations to Improve Productivity and Efficiency

In today's fiscal and political climate, it is hard to envision the creation of large new departments or agencies. Federal resources are stretched thin, and partisanship and political polarization are at high levels. In this situation, critics invariably question the need for another new federal agency. Even in the case of a large-scale financial meltdown, for example, it has been difficult for legislators to agree to create an independent consumer protection agency.

But lack of financial resources and the political will to create new offices is not the only thing retarding change. Public management now focuses less on creating new entities than on undertaking steps to improve the performance of existing agencies and departments. Observers believe the problem with government is less structural than behavioral. Instead of moving boxes around or constructing new agencies, we need to make current offices perform better.

Technology innovation represents a way to improve government service delivery and transparency, strengthen participation and collaboration, boost responsiveness, and save money. Technology can be used for online service delivery, public outreach, social networking, and civic engagement. We are early in this digital revolution, and we have an extraordinary opportunity to shift the manner in which government performs.

TRANSPARENCY AND ACCESS TO ONLINE SERVICES

When my colleagues and I began studying technology use in government ten years ago, only 22 percent of federal departments had online services.[7] That number has grown to almost 90 percent. Nearly two-thirds of Americans pay their income taxes online. Forty percent have gone online for government data, according to a 2010 survey by the Pew Internet and American Life Project.[8] One-third have renewed a driver's license or auto registration online. Twenty-three percent have participated in online debates about government policy, and 13 percent have read a government blog.

Many agencies have added novel features to their websites. For example, the Wyoming Supreme Court has an online database of its opinions. The state of West Virginia provides a live chat with customer support for website visitors. The Indiana portal helps visually impaired visitors by reading the web page aloud in English. The state of Michigan portal provides ten podcasts and seventy-two Really Simple Syndication (RSS) feeds in foreign languages such as Spanish and Arabic and offers some forms in languages such as Albanian, Chinese, French, Hmong, Korean, Polish, Russian, Serbo-Croatian, and Vietnamese. On the Wisconsin portal, a wizard helps users locate information on starting a business. These and other features have enhanced access to information and services.

At the federal level, the Obama administration has placed unparalleled amounts of federal data online at Data.gov. Each department chooses data sets for inclusion that help individuals and businesses. There is information on delays in airline flights, car safety ratings, crime statistics, small business loans, and business permits, among others. The user can download raw data, map information, or search for particular items of interest. Online resources lower the cost of information acquisition and make it easier for citizens and businesses to order services, data, and reports.

PARTICIPATION

Technology offers the potential to get more people involved with governance. When federal agencies consider new rules, there is a public comment period when individuals, groups, and businesses can give their reactions and make suggestions regarding proposed regulations.

In the past, the rule-making process was heavily dominated by industry and did not engage a broad diversity of perspectives. This limited the feedback federal officials received from outside sources. Now, many agencies allow electronic comments, and researchers have access to a broader range of views. Participation is no longer limited to those with a vested interest in the proposed regulation but is now open to consumer groups, interested experts, and ordinary citizens. In addition to lowering communication costs, technology makes it easier for federal officials to get comments from businesses, consumer groups, and advocacy organizations.

COLLABORATION

Through social networking sites, businesses and government can benefit from the ground-level view. For example, one of the difficulties in large organizations is moving information from those close to the scene of action to those who are making key decisions. Technology narrows the gap between information and decisionmaking, enabling leaders to make informed decisions.

The Pentagon has set up information-sharing systems that allow troops to convey their experiences directly to headquarters. Troops observe battle conditions and encounter obstacles that generals need to factor into their assessments. Those on the front lines have firsthand knowledge of everything from equipment breakdowns and food rations to enemy engagements. Befitting a department with a troop age structure generally running from eighteen to thirty, the Pentagon's social networking sites are among the best in the federal government.

RESPONSIVENESS

One of the most frequent complaints about government agencies is their lack of responsiveness. An individual requests information from a government agency and never gets a response. In earlier eras, that person would be upset but had few remedies to express discontent. Today, there are electronic metrics capable of tracking performance.

When I visited China a couple of years ago, I approached the airport passport control station and noticed an electronic device that allowed me to rate the performance of the passport official on a scale of 1 to 4. It was an ingenious way to monitor performance.

That kind of passport control tracker does not exist in the United States or many other nations, but the U.S. federal government has online surveys that allow people to rate their experience in requesting information or accessing online services at all federal agencies. Officials collect those responses and rate the agencies based on performance, citizen satisfaction, responsiveness, and ease of the transaction.

According to the latest American Customer Satisfaction Index of 250,000 government websites, undertaken by ForeSee Results, citizen satisfaction with public sector websites is high. In its recent report, the firm noted that "satisfaction with government websites climbed to an all-time high in 2009, and remains high for the most recent quarter. The E-Government Satisfaction Index registers a 75.1 on a 100-point scale in the first quarter of 2010." This represents the highest rating in the seven years ForeSee has been measuring public sentiments. In 2004, when the satisfaction index was first used, the overall rating was 70.9. It went to 71.9 in 2005, 73.5 in 2006, 73.4 in 2007, 72.4 in 2008, 73.6 in 2009.[9]

Top-performing agency portal sites include those of the General Services Administration (83 points out of 100), the National Aeronautics and Space Administration (NASA) (82 points), the National Park Service (82 points), the National Cancer Institute (81 points), the Federal Bureau of Investigation (FBI) (80 points), the Centers for Disease Control and Prevention (80 points), the Department of Defense (78 points), the Department of Homeland Security (77 points), the United States Mint (76 points), the Internal Revenue Service (73 points), the Food and Drug

Administration (68 points), and the Department of the Interior (65 points), among others.

These data help us improve public sector performance because they give information on the performance of government agencies. They do this in real time and allow administration authorities to determine what people like and dislike about their interactions with government.

SOCIAL NETWORKING

Social media represent a technology that is becoming more prevalent in American society.[10] Usage numbers demonstrate the rising popularity of Facebook, Twitter, and YouTube among the general public. Facebook estimates that it has more than 500 million active users worldwide and that 250 million log onto their home page every day. Twitter has more than 190 million visitors a month who send a total of 65 million messages a day. YouTube recently reached an audience of 135 million viewers.[11] Blogs have become common, and wikis are starting to be used to share information among interested groups of people.[12]

It is estimated that one-quarter of U.S. senators and representatives use some type of digital social media.[13] These and other social media tools offer the potential to advance governance through transparency, interactivity, and two-way communications between legislators and citizens.

Despite the rising popularity of digital communications, though, a number of barriers impede the achievement of desired benefits. Among the obstacles to bettering legislative policymaking and institutional performance are low usage levels, inadequate understanding of or incentives to use the interactive capacity of social media, a failure to use these tools to convey substantive information, and the employment of these media for one-way communications rather than general feedback.[14]

Reflecting an institution that has cumbersome procedures and has been slow to embrace new technologies, usage remains relatively low in many congressional offices.[15] There is little under-standing of the interactive capacity of social media, and those who use the tools generally fail to use them to convey substantive information to reporters or citizens.

These obstacles hinder the transparency, accountability, repre-sentation, and participation that are so important to our political system. There is considerable scholarly research demonstrating the importance of transparency in internal procedures and opera-tions. The U.S. House and Senate have rules and procedures dating back decades, and many of them are not fully transparent either to those within the chamber or interested external observers.[16]

Digital technology offers the potential to improve internal pro-cesses, making them more open and transparent. For example, Congress could use social media to communicate clearly and reg-ularly with its rank and file. The recent launch of Demcom.house. gov, a private intranet designed to facilitate two-way communica-tion on legislation and other matters among House Democrats, is a modest first step in this direction. Other applications, embracing both parties and committees, might help lessen the current polar-ization that exists between members of the majority and minority parties, increase substantive policy interactions among members and congressional staff, and improve the ability of journalists and other outside observers to follow what is happening in the legislative branch.

In addition, leadership accountability is crucial in democratic political systems.[17] It is difficult to hold legislators accountable when the traditional media provide little substantive information and the new media are underused. Numerous studies have found problems in the substantive content provided by newspapers and television stations. Most reporters focus much more on the horse race and the political game than on the policy substance necessary for accountability.[18]

New media offer the potential to improve the substantive con-tent of political communications. But if legislators use Facebook,

Twitter, blogs, wikis, and YouTube only for nonsubstantive information or scheduling activities (such as the member's travels), they miss an opportunity to provide information on their voting records, explanatory justifications for their activities, or other material that would help citizens judge legislative records.[19]

There also are problems of representation in the newly evolving system.[20] Social media are interactive technologies that allow members to communicate with constituents but also make it possible for citizens to provide feedback to legislators. Their interactive capacity offers tremendous hope for bridging the gap between representatives and constituents and supplying public feedback to members of Congress. Facebook, for example, enables legislators to post substantive information, explain votes, upload videos, and provide links to other Internet sites. It furthermore allows visitors to comment on a particular post or to provide a more detailed substantive evaluation of what the member has done.

This interactive capability makes it possible for members to receive real-time feedback from the general public that could enhance democratic representation. Feedback can be geared to specific proposals before Congress and can be dynamic over time. Voters often complain that members pay too much attention to special interest groups and ignore the general public interest. Digital technology combined with data-mining software could involve a broader range of participants and provide two-way communication with members of Congress. Yet so far, few members are employing these new tools to get regular feedback on what they are doing.

Social media can improve government efficiency. An example of this is the intelligence community's Intellipedia, whereby members of the community share information on common topics. The scholar Norman Ornstein has proposed a Legispedia, which would enable "staffers and Members to share insights and polish ideas about policies, witnesses, data sources and so on."[21] Such a digital entity would be especially useful for individual

24

congressional committees, or committees whose jurisdiction overlaps, to improve efficiency and encourage collaboration. The *National Journal* has launched a beta site, "LinkedIn for Congress," called 3121 (the capital switchboard phone extension number). It provides "biographical and job information for many employees" as well as regular news updates.[22] Through these and other types of innovations, the public sector can involve more people and become more effective at sharing information.

CONCLUSION

Digital technology and social media offer interesting possibilities to change the way the public sector functions and improve institutional performance. Similar to what has occurred in the private sector, it is conceivable that government could reinvent itself and produce greater efficiency and effectiveness. Technologies exist that enhance transparency, participation, collaboration, responsiveness, and social networking.

What is needed are steps to encourage public sector innovation so that we can take full advantage of the opportunities for innovation without suffering negative unintended consequences.[23] This involves a change in policies, the reengineering of organizations, and automation of routine tasks. These steps will allow government agencies to achieve new economies of scale through technology and improve the overall performance of the public sector. Despite public cynicism regarding government inefficiency, the public sector can perform faster, smarter, and more efficiently.

CHAPTER THREE

TRANSFORMING HEALTH CARE

HEALTH CARE TODAY IS DOMINATED by physicians, hospitals, the pharmaceutical industry, insurance companies, and government agencies. Patients navigating their health care move across a variety of health care providers, order prescription drugs from pharmacies, and seek reimbursement from either public or private insurance plans. They spend hours working out the best health care for themselves and their families. If they are fortunate to have good providers and effective follow-through, they receive high-quality health care.[1]

Imagine a different system in which, with the aid of the Internet, electronic medical records, cell phones, and personalized health care, the patient is in charge. The new health care system would be based on remote monitors, electronic medical records, social networking sites, video conferencing, Internet-based record keeping, and mobile health (mHealth). Mobile health incorporates e-mail reminders to take medicine, a Netflix-style mechanism by which customers can rate experiences with doctors and hospitals, and websites that make these ratings publicly available to employers and other patients.

People would monitor their own weight, blood pressure, pulse, and sugar levels and send test results via remote devices to health care providers. Patients would store their medical records online and have access to them from anywhere around the world. They would get personalized feedback by e-mail reminders when they gained weight, had an uptick on their cholesterol levels, did not take their medicine, or had high blood pressure. Social networking sites would provide discussion forums and the benefit of collective experience from other people suffering similar problems. Patients would take responsibility for their routine health care and rely on physicians and hospitals for more serious medical conditions.

This system is not a futuristic vision but is well within our grasp. Recent technological advances in communications, diagnosis, and treatment offer the potential to revolutionize the health care system and create better outcomes for patients and health care providers. To make optimal use of these advances, however, we must replace the incentives for volume-based health care that currently exist and replace them with incentives for good health outcomes and preventive medicine practices. With specific policy changes in reimbursement rates and insurance incentives, we can save money for the system as a whole and give individual consumers greater control over their medical care.[2]

A NEW HEALTH CARE SYSTEM

These incentives would entail significant changes in public and private insurance coverage as well as health care practices. These would include coverage that reimburses health care providers for mHealth care, remote monitoring, electronic communications with patients, electronic prescribing of medications, and downloading of medical tests to cell phones and other mobile devices; rewards for physicians who provide positive health outcomes for their patients, as opposed to paying them for the quantity of tests they have ordered; greater encouragement of preventive health care, good diet, and regular exercise activities; and creation of a preventive medicine fund that covers gym memberships, exercise equipment, flu shots, diet advice, smoking cessation programs, and substance abuse treatment.

Remote Monitoring Devices

A number of new remote monitors for various health care conditions put patients in charge of their own testing and keep them out of doctor's offices. For example, devices are now available that allow patients to measure certain vital signs (pulse, blood pressure, and glucose levels). AT&T has a new "device certification lab" that tracks health along high-speed broadband networks; results are electronically sent to a family physician, specialist, or electronic medical record, depending on the wishes of the patient. Zeo is marketing a monitor that measures brainwaves and rates the quality of sleep. Bodybugg has an armband calorie counter that charts the amount of energy burned through physical movements.

The Triage Wireless company has a wearable monitor that records vital signs and transmits them to physicians. The monitor records blood pressure on a continuous basis, providing regular information for health care providers. The Corventis corporation has a small sensor it calls PiiX that measures fluid status and respiration in runners. This helps people monitor their physical status during exercise. Intel offers a "magic carpet" product device that monitors physical movements. Geared for senior citizens, it tracks people as they walk on a mat and signals them if they are at risk of taking a fall.

It is crucial that diabetic patients monitor their blood glucose levels and gear their insulin intake to proper levels. In the old days, patients had to visit a doctor's lab or medical office, take a test, and wait days or weeks for results to be obtained. That process was expensive, time-consuming, and inconvenient for all involved and drove up the costs of health care.

However, remote monitoring devices allow people to record glucose levels at home instantaneously and electronically send the results to the appropriate health care provider. Patients are using

FDA-approved Gluco Phones that monitor and transmit glucose information to caregivers while also reminding patients when they need to take glucose tests. This puts people in charge of their own monitoring and keeps them out of their doctor's office until they need more detailed health care. It is estimated that more than 11 million Americans use home monitors for their glucose testing. Health authorities believe there are more than 24 million diabetics in the United States, and the disease is the seventh leading cause of death.

Tiny monitors with magnetic nanoparticles have been developed by researchers at the Massachusetts Institute of Technology to track the development of cancer tumors. Small particles the size of a rice grain are injected during biopsies. Through follow-up magnetic resonance imaging, doctors can measure whether these monitors clump with the tumor and grow in size. This allows them to get immediate feedback on the size of cancers and whether a specific therapy is working.

Cardiologist Steven Greenberg of St. Francis Hospital in Roslyn, New York, uses a wireless pacemaker made by St. Jude Medical connected to a home monitoring device to track heart rhythms and vital signs. Patient information is automatically transmitted to his medical office, which allows him to see which patient has an abnormal heart beat and therefore is in need of immediate treatment. Greenberg feels this enables him to "stay a step ahead of potentially life-threatening problems."[3]

Personalized Reminders and Feedback

One of the biggest problems in medical treatment is patients' forgetting to take their prescription drugs. It is estimated that only 50 percent of patients with chronic conditions take their medication as prescribed. Some forget to take the drug; others fail to take it at the time or dosage set by their physician. This means that half of the benefit of prescription drugs is lost through human error, at a cost of billions in poor health outcomes.[4]

Digital technology has the potential to help with this and other communications problems. Patients no longer need to visit a doctor's office to be reminded to take their medicine. They can get personal reminders by e-mail, automated phone calls, or text messages. One enterprising physician named David Green, of Cape Town, South Africa, noticed that his patients did not always take the prescribed Rifafol medicine for their tuberculosis. He knew that unless the pill were taken on a consistent basis, it would have little effect. Doctor Green set up a text-messaging service called On-Cue Compliance for his patients that sent them a daily short-message service (SMS) in English, Afrikaans, or Xhosa. Over the six-month course of treatment, his service would send a message at a predetermined time each day reminding them to take their Rifafol.[5]

In the United States, Dynamed Solutions manufactures HealtheTrax software that reminds patients to take medications, set up appointments, and track compliance with medical instructions. This and other types of "virtual health assistants" are particularly helpful with those suffering from chronic illness, who need to keep close track of their medical condition and stay in touch with their caregivers. The software is integrated with electronic medical records and can store information in patients' personal records.

For young people on the go, remembering to take medication is one of the biggest challenges. Physicians at Children's Hospital Medical Center in Cincinnati send teenagers text messages reminding them to take their asthma medication. Researchers have found that text reminders are an effective means to compliance with therapy and help teenagers develop good self-care habits.

A company called Proteus Biomedical makes a tiny digestible chip that is swallowed along with a prescription drug and notifies health care providers that patients have taken their medication. Using a sensing device, the chip electronically transmits that information to physicians, who then know for sure that the individual is following the prescribed course of treatment. It is especially helpful with patients suffering memory loss. Most patients like getting personalized reminders from their medical providers. One person indicated that these messages "keep you informed and mean you never forget to take your drugs."[6]

In general, Americans say they would like to employ digital technologies in their medical care. For example, in one national survey 77 percent said they would appreciate getting reminders by e-mail from their doctors when they are due for a visit, 75 percent want the ability to schedule a doctor's visit by the Internet, 74 percent would like to use e-mail to communicate directly with their doctor, 67 percent would like to receive the results of diagnostic tests by e-mail, 64 percent want access to electronic medical records to capture information, and 57 percent would like to use a home monitoring device that allows them to e-mail blood pressure readings to their doctor's office.[7]

Mobile Smartphones

Cell phones and other mobile devices have gotten smarter and faster. Smartphones such as Apple's iPhone, Research in Motion's BlackBerry, Nokia's E71, and Hewlett Packard's Palm Pre offer advanced features such as mobile e-mail, web browsing, and wireless communications. The sophistication of these devices has spawned a variety of new medical applications that help doctors and patients stay in touch and monitor health care needs.

For example, Sprint has a mobile application that allows physicians to get test results on their mobile device. They can look at blood pressure records over time, see an electrocardiogram, or monitor a fetal heart rate. AirStrip Technologies markets an application that makes it possible for obstetricians to remotely monitor the heart rates of fetuses and expectant mothers. This allows them to detect conditions that are placing either at risk.

These applications increase the efficiency of doctors by enabling them to judge a patient's condition without being in his or her physical presence. With digital technology, people can overcome the limitations of geography in health care and access information at a distance. For example, patients can get a second opinion remotely, by sending that doctor relevant medical tests. If a personal conference is required, doctors can use video conferencing to speak to patients located in another city or state.

Electronic Medical Records

In the past, medical records were stored on paper in dusty files. This required a lot of storage space, and records were hard to access from a distance. Physicians did not have records at their fingertips, and patients had little control over their own medical records.

Today, medical information can be stored on Internet sites or portable devices. Microsoft Health Vault and Google Health maintain online medical records for patients. Here patients can record and chart their cholesterol levels, blood pressure, glucose readings, weight, pulse, and other health data. Permission to view various levels of this information can be given to primary care physicians or medical specialists. In cases of emergency, vital records can be made available to the treating doctors or nurses.

Internet Information

There has been an explosion of websites with detailed medical information. Websites such as WebMD.com, MedlinePlus. gov, MerckSource.com, HealthFinder.gov, and MayoClinic.com answer questions and provide links to discussion groups about particular illnesses. In states such as Massachusetts, California, New York, and Michigan, consumers can visit health department sites and compare quality performance data on provider care programs. Nationally, the U.S. Department of Health and Human Services has a website (www.hospitalcompare.hhs.gov) that evaluates 2,500 hospitals on mortality rates, room cleanliness, call button responses, and patients' judgments of quality of care.[8]

According to a 2007 Harris Interactive survey, the most common Internet searches occur in regard to specific diseases. Of those who went online, 64 percent said they searched for information on particular illnesses, 51 percent looked for certain medical treatments, 49 percent surfed for material on diet and nutrition, 44 percent named exercise, 37 percent sought advice on medical drugs, and 29 percent looked for particular doctors or hospitals.[9]

This information had a positive impact on many people's medical decisions. In the Harris survey, 58 percent indicated that online material affected their health care decisions, 55 percent said the information changed their health care approach, and 54 percent claimed the electronic resources made them ask new questions of their medical personnel. When asked how these materials made them feel, 74 percent said they felt reassured and 56 percent felt more confident, but 25 percent indicated they were overwhelmed by the amount of information online, 18 percent were confused, and 10 percent frightened by the medical material they uncovered.[10]

From these findings, it is clear that most people's experience researching online helped them learn more about illness and treatments. But others had difficulty dealing with this new online world. They did not feel comfortable searching for online information and got confused or overwhelmed by what they found at medical websites.

Social Networking for Medical Care

Social networking sites allow sufferers of chronic conditions to share information about their disease and its treatment. For example, a network developed by the company Patients Like Me has 23,000 patients who have signed up to share information regarding five different illnesses: mood disorders, Parkinson's, multiple sclerosis, HIV/AIDS, and Lou Gehrig disease. Users of the site describe their symptoms, discuss various therapies, and talk about what therapies have worked for them and what have not. The site not only serves as a vital support group for these serious illnesses, but it also promotes better understanding through the detailed case histories based on personal experiences.

It often takes years for patients, physicians, and medical researchers to get definitive results regarding the assessment of drugs and medical therapies. Clinical trials are expensive and time consuming and involve randomized assignment to various groups. Results sometimes are unclear, and it is hard to recruit sufficient subjects to participate in the evaluations. While it is important to maintain rigorous approaches to medical research, it is helpful to take advantage of new techniques for getting feedback.

One new technology draws on crowd sourcing for feedback regarding medical care and side effects of treatment. Crowd sourcing takes advantage of the collective experience of large groups of people. It allows a variety of people to comment on and post their experiences with specific treatments. This helps others compare data and see information on the effectiveness of therapies.

Dr. Amy Farber has developed an online resource, at LAMsight. org, that encourages people suffering from the lymphangioleiomyomatosis (LAM) lung disease to share their symptoms and treatment experiences. Web operators take this patient-provided information and compile online databases that are used by researchers to find out which treatments have been effective and which drugs generate unwelcome side effects. This is particularly helpful in the case of rare illnesses, where it is hard to generate the patient numbers required for clinical trial.[11] While large clinical trials with randomized assignment clearly need to remain central to drug assessment, digital technology that helps providers and researchers identify worrisome trends represents an additional way to gain useful feedback.

Consumer Evaluations of Health Care Providers

A big challenge in contemporary health care is lack of information among patients about the quality of physician and hospital care. There is some outcome-based information on how many mammograms or other medical tests are performed by various facilities but few assessments of the quality of care received from specific providers.

Digital technology has the potential to empower the consumer voice in health care and to share patient assessments of doctor performance. This is already being done in the entertainment area, for example, the commercial company Netflix has devised a system by which film watchers order movies for home viewing. Upon returning the movie to the company, customers receive an automatic e-mail asking them to rate the movie on a five-point scale. This information is anonymously aggregated and publicly available to other consumers so they can see which movies receive the highest ratings in various categories.

It is possible to create a similar system for rating physicians, hospitals, and other health care providers. Following physician visits, consumers would be offered an e-mail form allowing them to rate different dimensions of their medical treatment, from timeliness and personal attentiveness of the staff to the doctor's level of knowledge and the overall consumer's satisfaction with the visit. These quality measures would be aggregated and become accessible at a public website so others could see the quality assessments.

Consumer Reports offers an online hospital rating service of 3,400 facilities based on the national government's Hospital Consumer Assessments of Healthcare Providers and Systems Survey.[12] This site ranks hospitals on indicators such as patient experiences, room cleanliness, and noise level, among other things. Preliminary results suggest that caregivers communicate poorly about new medications and that one-third of hospital staff is inattentive to patients.

REQUIRED CHANGES

There is little doubt that the technology for customer-driven health care is already available. These tools cannot be implemented, however, without incentives for patients and health care providers that reward good behavior and good health outcomes over volume health tests.

Reimbursement Changes in Insurance Coverage

Few Americans currently take advantage of health information technologies. The 2007 Harris survey mentioned earlier asked 2,624 adults across the nation about their use of certain electronic tools in their health care. Only a small number indicated they currently used any of these electronic tools. Four percent got reminders by e-mail from their doctors when they were due for a visit, 4 percent used e-mail to communicate directly with their doctor, 3 percent scheduled doctor visits through the Internet, 2 percent received the results of diagnostic tests by e-mail, 2 percent had access to electronic medical records, and 2 percent relied on home monitoring devices that allowed them to e-mail blood pressure readings directly to their doctor's office.

Part of the problem is that neither public nor private insurance plans cover most mHealth, digital communications, or wellness programs. Physicians, for example, often are not reimbursed for e-mail or phone consultations. We need policy changes that encourage high-quality medical care and make it possible for health providers to be adequately reimbursed for the care they provide.

We should reward good behavior and reduce the incentives for wasteful or unnecessary treatment. For example, rather than paying providers for each test or office visit they provide, insurance companies could provide an annual sum to family doctors based on the person's age and health status. This would encourage both patients and providers to monitor overall health and make decisions on where best to allocate health care dollars. According to a July 26, 2009, article in the *Washington Post,* "Pilot projects have shown that paying a medical team for total care—monitoring blood-sugar levels, giving eye and foot exams—rather than paying for each visit to an ophthalmologist or podiatrist is better for the patient and costs less."[13]

The Geisinger Medical Center, in northeastern Pennsylvania, tested a "medical home" initiative among Medicare patients and found that the program resulted in an 8 percent drop in hospital admissions and a 4 percent reduction in overall health costs over the first year. In this concept, patients are assigned a family physician who acts like a personal health coach. This coach oversees a group of providers who monitor patients' medical condition and use e-mails and text messages to encourage them to lose weight, stick to healthy diets, get exercise, and seek relevant care when their status deteriorates.

A Focus on Positive Health Outcomes

Doctors and hospitals make money by ordering tests. Virtually all the current incentives encourage this volume-based health care. With the fee-for-service basis of contemporary health care, there is a financial incentive for health providers to see a lot of patients and provide treatment, some of which may be duplicative or unnecessary.

In this system, there is little attention to health outcomes, either for individual health care providers or the medical system as a whole. Doctors are not rewarded for healthy patients or preventive medical care. Indeed, one of the challenges in the current system is the lack of performance data on health outcomes. The federal government collects statistics by city and state on causes of death, numbers of procedures, and other such information but little data on the state of patients' health.

Outcome information for specific doctors or other health care providers is also lacking. This makes it difficult to judge quality or create incentives for healthy outcomes. We need to supplement the current fee-for-service with a fee-for-performance that links insurance reimbursements to positive outcomes. Doctors whose patients remain healthy should receive a bonus and should be encouraged to continue preventive medical care.

Rewards for Good Behavior

Drivers who do not have accidents or are not cited for speeding or other traffic violations earn a 10 percent discount on their insurance. The program protects drivers by rewarding safe driving and is cost-effective for car insurance companies because safe drivers have fewer accidents and therefore cost the company less in accident repair reimbursements. The health care industry could benefit from similar models.

Americans eat too much, get too little exercise, and have diets that are too fatty. The result is an obesity epidemic that will push health care costs higher in the future and limit people's quality of life. According to the American Obesity Association, more than 30 percent of children today are overweight. This ticking time bomb threatens to explode and have dramatic consequences for national health care spending.

Insurance companies and government programs should offer "good health" rewards to patients and physicians. Health programs could provide a preventive medicine fund, for example, that would reimburse people for regular exercise, good health practices, flu shots, diet advice, and smoking, alcohol, or drug cessation programs. This would encourage patients to lead healthy lifestyles.

Some writers such as David Leonhardt of the *New York Times* have gone so far as to propose a "fat tax" that would charge higher health insurance premiums for individuals with a high body-mass index.[14] This would penalize those who are overweight and perhaps provide financial incentives for them to lose weight. And it would reward those with healthy lifestyles by offering 20 to 30 percent discounts on insurance premiums.

After the Safeway company instituted a Healthy Measures program of cholesterol screenings, blood pressure measurements, and weight loss initiatives, its health costs dropped by 13 percent. More than three-quarters of its employees enrolled in the program, and they saved 20 percent on their individual insurance premiums. Pitney Bowes provides $100 gift cards to employees who enrolled in health courses. Senators Max Baucus and Tom Harkin have proposed employer tax credits for company programs that provide health screenings and lifestyle counseling.

SAVING MONEY AND LEADING HEALTHIER LIVES

The ultimate goal of health care policy changes is to save money and encourage people to lead healthier lives. The United States spends $6,102 per capita a year on health care, much more than the $3,165 spent by Canada, $3,159 by France, and $2,083 by the United Kingdom. Yet America ranks forty-second among developed nations in life expectancy. Our average life expectancy of 77.9 years falls well below that of Andorra, the Cayman Islands, and most European countries.[15] We spend a higher percentage of gross domestic product (GDP) on health care than most other nations but get weaker results in terms of medical well-being.

With America's health care system now costing $2.4 trillion a year, we can no longer afford delays in making needed changes. As Peter Neupert of Microsoft's Health Solutions Group has written, "Let consumers do some of the work that expensive health-care professionals shouldn't be doing anymore. In the past 10 years, technology has removed travel agents, bank tellers and so on from the middleman position. Online systems, such as Kaiser Permanente's, have increased patient satisfaction and allowed the work of expensive professionals to be replaced."[16]

One of the reasons America spends more money per patient than other countries but gets weaker results is our low usage of health information technology. Only 15 percent of the 560,000 doctors in America use digital technology to order medication for their patients.[17] According to health experts, digital technology would save money and "make transactions more efficient, reduce medication errors and entice doctors to prescribe less expensive drugs."

An analysis undertaken by the economist Robert Litan finds that remote monitoring technologies could save as much as $197 billion over the next twenty-five years.[18] Cost savings are especially available in the chronic disease areas of congestive heart failure, pulmonary disease, diabetes, and skin ulcers.

With round-the-clock monitoring and electronic data transmission to caregivers, remote devices could speed up the treatment of patients requiring medical intervention. Rather than having to wait for a patient to discover there is a problem, monitors could identify deteriorating conditions in real time.

A PricewaterhouseCoopers Health Research Institute study has found that $210 billion is wasted through "defensive medicine—doctors ordering tests or procedures not based on need but concern over liability or increasing their income."[19] Other examples of wasteful spending include inefficient claims processing ($210 billion), disregard of doctor's orders ($100 billion), ineffective use of technology ($88 billion), hospital readmissions ($25 billion), medical errors ($17 billion), unnecessary emergency room visits ($14 billion), and hospital-acquired infections ($3 billion).

Better use of digital and mobile technology could help on each of these fronts. Electronic medical records would reduce duplicate tests because various physicians would have easy access to records of past procedures and their results. Automated processing of medical reimbursements would save time and money. Noncompliance in taking medicine at prescribed times and levels could be improved through remote monitoring and digital tracking. Unnecessary emergency room visits, hospital infections, and medical errors could be reduced through personalized medicine that employs video conferencing and treatment facilities other than hospitals.

Some observers worry that electronic consultations will depersonalize health care. The social medicine expert Helen Hughes Evans, for example, argues that "technology has stripped medicine of its humanistic qualities" and that doctors rely too heavily on high-tech instruments.[20] Rather than advancing the quality of patient care, she feels, digital medicine has harmed the intimacy of clinician-patient relations and undermined the personal touch within health care.

In a review of telemedicine research, though, Edward Alan Miller finds that 80 percent of medical studies show a favorable impact of digitally mediated contact on provider-patient relations.[21] These kinds of technologies facilitate access for some individuals and expand the network of health care providers that are available to patients.

Digital communications allow people with rare diseases to find others who suffer from the same disorder and to learn from their experience. These systems also allow patients to take advantage of specialists in states or even countries far removed from their own. Personalized medicine can increase resources for self-care, enhance emotional support through electronic support groups, and improve knowledge regarding specialized medical problems.

PART TWO

PLATFORM

CHAPTER FOUR

CLOUD COMPUTING

THE U.S. FEDERAL GOVERNMENT SPENDS nearly $76 billion each year on information technology (IT), $20 billion of which is devoted to hardware, software, and file servers.[1] Traditionally, computing services have been delivered through desktops or laptops operated by proprietary software.[2] But cloud computing, a new advance in information technology, makes it possible for public and private sector agencies alike to access software, services, and data storage through remote file servers. With the number of federal data centers having skyrocketed from 493 to 1,200 over the past decade, it is time to more seriously consider whether money can be saved through greater reliance on this new technology.[3]

As Jeffrey Rayport and Andrew Heyward note, cloud computing has the potential to produce "an explosion in creativity, diversity, and democratization predicated on creating ubiquitous access to high-powered computing resources."[4] By freeing users from reliance on desktop computers and specific geographic locations, clouds computing revolutionizes the communication tasks.

Because they are based on a network of remote file servers distant from the user, clouds enable organizations to scale up or down to the level of needed service so that users optimize their needed capacity. Fifty-eight percent of information technology executives in the private sector anticipate that "cloud computing will cause a radical shift in IT and 47 percent say they're already using it or actively researching it."[5] Cloud computing clearly is becoming the platform of choice for many Internet users.

CHARACTERISTICS OF CLOUD COMPUTING

The National Institute of Standards and Technology distinguishes cloud characteristics, delivery model, and deployment method.[6] It lists five key features of cloud computing: on-demand self-service, ubiquitous network access, location-independent resource pooling, rapid elasticity, and measured service.[7] Cloud computing can take the form of software-as-a-service (running specific applications through a cloud), platform-as-a-service (using a suite of applications, programming languages, and user tools), or infrastructure-as-a-service (relying on remote data storage networks). Deployment depends on whether the cloud is a private, community, public, or hybrid one. Private clouds are operated for a specific organization, for example, whereas community clouds are shared by a number of organizations. Public clouds are available to the general public or large groups of agencies, whereas hybrid clouds combine public and private elements in the same data center. The dynamics of operations and cost savings depend on the type of cloud being envisioned.

ESTIMATING CLOUD SAVINGS

There is a wide range of cloud saving estimates from various sources, some of which are much more optimistic than others. Former Office of Budget and Management official Mark Forman estimates that migration to the cloud will save 90 to 99 percent of IT operating costs.[8] A report by Ted Alford and Gwen Morton of Booz Allen Hamilton concludes that government agencies moving to public or private clouds can save from 50 to 67 percent. Rajen Sheth of Google projects cost savings of 67 percent for moving e-mail to the cloud.[9] MeriTalk puts the overall cost savings at around 39 percent, based on projected economies from open source, virtualization, and cloud service delivery.[10]

However, a report by McKinsey analyst William Forrest disputes these cost savings.[11] He argues that cloud migrations would yield few savings and would actually cost 144 percent more than current expenditures. He bases this analysis on a comparison of file server costs through a conventional data center with those of Amazon's Elastic Compute Cloud. Whereas the regular storage mechanism would cost $150 a month for three GHz dual-core Xeon servers ($107 for labor and $43 in other costs), Forrest estimates Amazon's cloud storage costs at $366 a month ($270 for labor and $93 in other costs).[12]

The wide variation in estimates of return on investment demonstrates that there is considerable uncertainty in projected cost savings and a need for studies that look at actual expenditures. Uncertainty is not surprising, given the many factors that go into cost estimation. One factor is the scope and timing of the migration. It matters how extensive the migration is and whether the cloud deployment focuses on applications, service delivery, or platform storage. The bigger the migration, the higher the expected transition costs and the higher the labor costs involved. Simple migrations offer greater potential for cost savings than complex moves because of the labor costs, time, and expense of the migration.

The type of cloud being used also affects the cost savings that will be generated. Reliance on public versus private clouds makes a big difference. Alford and Morton, for example, find that an agency needing 1,000 file servers will spend $22.5 million for storage on a public cloud, $28.8 million on a hybrid, and $31.1 million on a private cloud limited to the agency. This means that a private cloud typically would cost about 40 percent more than a public cloud owing to greater security needs and lower storage usage rates.

In comparing studies, it is clear that the efficiency of capacity usage is another determinant of cost savings. The higher the capacity usage in an agency, the more likely will cost savings be because after a migration the agency can reduce the number of file servers it needs. Alford and Morton have found that many government data centers average 12 percent usage, meaning that agencies use only 12 percent of their storage space. If migration to a cloud increased usage to 60 percent, that would translate directly into dollar savings.

The level of privacy and security protection is another variable in cloud migrations for government agencies. Organizations that have sensitive or classified information obviously require greater safeguards, in terms of both monitoring and firewalls. This subsequently affects the cost of cloud storage and service delivery.

One final significant factor determining the level of cost economies is the extent of labor savings, that is whether migration to the cloud enables an agency to reduce personnel. Organizations that are able to downsize their IT departments based on cloud migration will see greater cost savings. For example, Forrest argues that agencies could save 15 percent on labor costs by moving to a cloud. But these types of savings are possible only if agencies actually cut personnel through cloud computing.[13] In general, staff reductions are politically and organizationally difficult for government agencies.

DATA AND METHODS

The data for this project come from a series of case studies involving government agencies that have moved specific applications from local to remote file servers. Specific analysis involved the cities of Los Angeles, Washington, Miami, and Carlsbad, California, as well as the U.S. State Department, the National Aeronautics and Space Administration, and the U.S. Air Force. I used interviews, media coverage, case materials, and documentary research to determine each agency's motivation in moving to the cloud, its cost structure before the move (if available), costs following the transition, cost savings in hardware, software, and personnel, and any difficulties experienced during the migration. Not every organization surveyed was able to provide complete information in each category, but I report all the available data that I was able to obtain.

E-MAIL SERVICE IN THE CITY OF LOS ANGELES

In 2009 the city of Los Angeles voted to move e-mail service for its 30,000 employees from Novell's GroupWise onto cloud file servers operated by Google. The $7.5 million contract provided five years of e-mail services for city employees at an average cost of $50 a year per employee. During the city council's consideration of the bill, critics worried about security and reliability, especially for law enforcement agencies such as the Police Department.[14] Google promised to store city data on its secure GovCloud platforms that are maintained within the continental United States and operated by individuals with high-level security clearances who have passed FBI fingerprint checks. The company also agreed to provide financial credits to the city if the system is down beyond service levels agreed to in the e-mail contract.

An analysis undertaken by the city administrative officer Miguel Santana for the city council has found that the five-year costs of running the Google system would be $17,556,484, which was 23.6 percent less than the $22,996,242 for operating Group-Wise during that same period.[15] The Google estimate comprised three pieces:

—$10,664,445 for system applications, implementation, a required Internet update, two file servers, and four personnel positions necessary to run the system;

—$907,913 for GroupWise e-mail licenses and software for 2009–10 during the transition to Google;

—$5,984,126 for Microsoft licenses for selected employees who would continue to use Office software during the five-year period.

If the city had chosen to stay with its current GroupWise e-mail system, the costs would have been significantly more:

—$15,459,438 for GroupWise licenses, upgrades, system applications, ninety file servers, and thirteen staff positions; and

—$7,536,804 for Microsoft Office licenses.

As far as personnel savings, the city would need nine fewer people in its information technology department as a result of the transition to the Google cloud. The Los Angeles Information Technology Agency initially requested that the nine people whose jobs no longer were needed for e-mail operations be retained for use on other technology projects. But the city administrator turned down that request, and the agency agreed to eliminate the nine jobs over a period of time.

For the eighty-eight file servers no longer needed to support e-mail, the Information Technology Agency decided to redirect sixty to replace obsolete servers in city government. Currently, the City of Los Angeles has 245 file servers that are five to ten years old and therefore in need of replacement. Typically, the city spends $1 million a year to purchase fifty-two new and replacement file servers. The IT agency agreed to forgo new file server purchases until the e-mail servers were deployed in place of obsolete machines.

E-MAIL SERVICE IN WASHINGTON, D.C.

In 2008 the city government of Washington, D.C., shifted many of its 38,000 employee e-mail services across eighty-six agencies to the cloud.[16] When he came into office, the city's chief technology officer, Vivek Kundra, found that "85 percent of school computers had viruses" and that the city's fiber-optic lines were costing $6.3 million more than budgeted.[17] He decided drastic action was required to improve service delivery and save money. The new e-mail contract was not an exclusive arrangement, as most workers were not required to shift to the Google system. A number continued to rely on Microsoft Outlook.

Government officials also employed a cloud approach that "plots the locations of construction projects and broken parking meters, among other things, on Google Maps, so residents can see how many potholes are scheduled to get filled on their street or how many computers a neighborhood school received this year."[18] Google Docs and Google Spreadsheets were used to store work flow data on city projects and employee performance information. Analysts found that the migration saved 48 percent on e-mail expenditures. Costs for Google Apps over the previous software were reduced from $96 to $50 a year per user.[19]

E-MAIL SERVICE IN CARLSBAD, CALIFORNIA

The City of Carlsbad decided in 2008 to move its 1,100 employees from Novell GroupWise onto a Microsoft Online Services e-mail and web conferencing solution. The service was implemented on a per-user, per-month payment basis, and the city estimates that it saved 40 percent a year compared with its earlier, in-house system. This included the cost of file servers, software upgrades, and staff training. For eighty mobile workers who did not have offices, the City saved $4,800 on e-mail licensing costs, an average of $60 per employee.[20]

311 MANAGEMENT IN MIAMI, FLORIDA

The City of Miami chose a Microsoft Windows Azure platform in 2009 for its service hosting and mapping technology. Before this decision, Miami hosted its own file servers over a storage area network and anticipated it would need terabytes of storage over a five-year period. However, when it discovered after three years that it was using twenty-seven terabytes of space, city officials moved to the cloud to track its 311 services to residents on potholes, illegal dumping, and missed garbage collection.

Azure offered a pay-as-you-go mechanism that worked for the city, given its tight budget and 18 percent drop in IT spending authority during the recession. The city combined Azure with a product from Imager Software known as MapDotNet UX, which made possible the visual display of information through geospatial mapping.

Because of the cloud storage, the city was able to drop much of the need to host and maintain its own physical file servers. City officials estimated that it saved 75 percent in the first year between hardware, software, and staff efficiencies.[21] Since this is a new system, cost figures for out-years were not available from Miami administrators.

BUDGET INFORMATION FOR THE NONPROLIFERATION AND DISARMAMENT FUND

The U.S. State Department's Nonproliferation and Disarmament Fund needed an application that would make budget information on nonproliferation issues available to out-of-the-office program managers around the world operating 24/7. When they were traveling, these officials would not have regular access to desktop computers or management budgeting tools. They needed a ubiquitous, multiplatform application that would let them know how much they had spent on particular nonproliferation issues and what money was left to pursue new program opportunities.

In 2008 the agency contracted with SalesForce.com for a new application that would provide detailed budget information from any Internet service anywhere in the world. The feature cost $1,426,691, including software, staff time, operations, and implementation. Executives estimated that this application cost them one-quarter what it would have, had development been undertaken in-house. Cost savings included reduced personnel needs and savings in development time as well as savings from file servers that did not have to be purchased.[22]

INFRASTRUCTURE PLATFORM FOR NASA

The National Aeronautics and Space Administration (NASA) has pioneered a new cloud system, known as Nebula, that is used for mission support, public education, and data communications and storage. It relies on Eucalyptus, an open-source cloud platform developed at the University of California at Santa Barbara.[23]

Nebula is mostly used for internal projects at NASA's Ames Research Center in Mountain View, California, but the agency also leases some storage space to the U.S. Office of Management and Budget so that it is fully used. Initial funding for the project was $2 million, less than the cost if NASA had to develop its own software and data file servers. The organization decided to develop its own cloud instead of using commercial services because the latter did not have the bandwidth required for NASA missions or were not yet compliant with the security specifications of the Federal Information Security Management Act.

According to Gretchen Curtis, director of communications for the Nebula cloud computing platform, Nebula has pioneered a new model of government cloud computing.[24] When state and local organizations talk about moving to "the cloud," they are most often referring to software-as-a-service functions, such as e-mail and other end-user services delivered over the Internet. "Here at NASA," she says, "we're talking less about software as a service, and more about infrastructure as a service (IaaS). . . and platform as a service (PaaS)."

As a science organization, NASA processes and stores a large amount of data. Consequently, the agency's need for storage and computing resources is very high. Currently, Nebula provides infrastructure-as-a-service, is moving toward platform-as-a-service, and eventually plans to start adding software-as-a-service. "We have plans to make Nebula available to a much wider audience," reports Curtis. "We have a handful of select beta account users, and we're going to be adding to that list."

Cloud computing helps NASA scale up or down given levels of scientific needs and public interest. Curtis notes that "[NASA] needed really powerful computing for short amounts of time, so we built Nebula to kind of cater to the scientific community." For example, in the days leading up the LCROSS event (the search for water on the moon), the agency anticipated that there would be a huge spike in traffic, which would require a great deal of computing power that would then die down, and so require less computing power. Organizations and agencies can realize greatest efficiencies by thinking about what kind of needs they are trying to address and then using the cloud accordingly to meet that need.

Specific cost savings are hard to compute because the project did not exist before Nebula. Computational savings vary on a per project basis, depending on how much computing power or storage is needed and for how long. One example is a Nebula project processing high-resolution images of the moon and Mars, for use in a worldwide mapping project. Data were sent back to Earth from satellites in space, processed, and sent to Microsoft for placement on a 3-D map of the world. Such a project requires powerful computing.

In the absence of Nebula, NASA would have had to procure a lot of new infrastructure for the computing and storage resources to be powerful enough to handle that amount of information. The procurement process would have involved having to justify needs for the hardware, shopping around, waiting for paperwork to go through, waiting for approval, and then waiting for the new machines to arrive. Once the machines did arrive, the agency would have had to hire system administrators, who would then spend a month or two configuring the machines for use; even then, the machines would not be used all the time, only when new data were coming in. Finally, once the project concluded, the group would end up with a bunch of useless hardware.

By employing Nebula to take care of computing and storage, NASA was able to provision virtual machines and get up and running right away. As Curtis puts it, "It was a matter of minutes, versus months." In addition, once the project was completed, the resources in the Nebula cloud could be used by others. Curtis estimates that the group saved four to five months of going through the procurement process, the many man hours involved in that, a month and a half of full-time work for a couple of systems administrators, and that major financial costs would not recur. That, she feels, represents a major cost savings.

DATA STORAGE FOR THE U.S. AIR FORCE

The U.S. Air Force 45th Space Wing is responsible for launching and tracking unmanned space vehicles from the Cape Canaveral Air Force Station and employs more than 10,000 workers. The Space Wing had sixty distinct file servers but found that it used only 10 percent of central processing unit capacity and 60 percent of random access memory space. Information technology supervisors found that low usage levels cost hundreds of thousands of dollars each year, so they decided to replace their old file servers with four servers running WMware ESX. The base stores data at two sites (Cape Canaveral and Patrick Air Force Base) and uses an Internet cloud to link the data centers.

Commanders estimate that they save $180,000 a year in computing costs. This includes $104,000 in hardware costs, $30,000 in power to cool what used to be sixty file servers, $28,000 in maintenance costs, and $18,000 in other expenses.[25] In addition, the unit no longer has to devote financial resources each year to buying new hardware or deploying new software. There were no estimated personnel savings.

MOVING FORWARD

Right now, cloud computing represents a relatively small amount of federal IT spending. In 2008, for example, only $277 million of the federal government's IT budget was devoted to cloud computing—a tiny fraction of its $20 billion in hardware and software expenditures.[26]

Currently, the federal government provides cloud solutions through www.Apps.gov. This site allows agency officials to purchase cloud computing services in the areas of business, productivity, and social media applications. Among the business applications are those for asset management, business processes, dashboard, data management, geographic information, surveys, and travel. Productivity applications include video conferencing, office tools, project management scheduling, and work flow. Social media possibilities include search tools, blogs, videos, and contests. Most of these fall within the Federal Information Security Management Act certification for low-risk applications.

Since there are few options for moderate-risk applications, which require greater security standards, the U.S. General Services Administration has withdrawn its blanket purchase agreement and is drawing up new procurement documents for infrastructure-as-a-service. This will allow the government to take on more sensitive cloud needs from agencies and departments having more advanced security requirements.[27]

A General Services Administration information technology initiative is completing a benchmarking study of possible cloud computing savings across the federal government. The agency is seeking to determine possible economies of scale resulting from cloud computing. It has devised an ambitious timetable for implementing cloud solutions in the federal government and collecting data on impact and performance. Eventually, the federal government aims to meet high-risk security needs for agencies such as the Department of Defense. It plans to roll out these applications shortly for software, platform, and infrastructure needs.[28]

The agency has created a federal cloud computing executive group composed of federal executives and chief information officers. It has developed a technical framework for federal cloud computing and issued a data call to federal agencies on cloud-related issues. These types of evaluations should be undertaken regularly so leaders and citizens can understand progress being made. The assessments should include the amount spent on technology, security and privacy protections, and agency adoption and innovation activities.

Another factor that is important to long-term cloud cost savings is some agreement in national laws across borders.[29] Countries have different rules or norms on cloud computing, privacy, data storage, security processes, and personnel training. Realizing the full efficiency of cloud computing is hard when laws are inconsistent or contradictory. Regulations on cross-border transactions via the cloud should be clarified and harmonized when possible to facilitate innovation and get the greatest economies of scale from this new technology.

Government information technology is subject to a variety of rules, regulations, and procurement policies. Computing is treated differently depending on whether the platform is based on desktops, laptops, mobile devices, or remote file server—the cloud. There are differences between the executive, legislative, and judicial branches of government as well as in the level of privacy and security expected for various applications.

Some people perceive higher security on desktop or laptop computers and lower security with the cloud computing because with the cloud, information is stored remotely through third-party commercial providers. In reality, though, there are serious security threats to all electronic information regardless of platform, and cloud server service providers often take security more seriously than mass consumers or government officials using weak passwords on their local computers.

NEEDED CHANGES

A number of policy changes in federal practices, policies, and procedures would improve the efficiency and effectiveness of federal IT policy. These might include improving efficiencies in data storage, expanding the use of communication tools in congressional offices, updating the federal courts' use of computing technology, centralizing contracting and certification processes, improving efficiency in procurements of goods and services by expanding the federal Apps.gov site, seeking international agreements on cloud computing rules and norms across borders, ensuring portability of information across platforms, and communicating the advantages of information transferred via the cloud to the public.

Data Consolidation

The financial cost and energy requirements of maintaining federal data centers is enormous. Most data centers function at low-level storage capacity to ensure that agencies never run out of disk space. Some studies have estimated federal storage usage running as low as 7 percent.[30] Others have found storage capacity averaging from 7 to 15 percent.[31]

Cloud storage and service usage provide greater efficiency because users pay only for the server space they actually use. It is easy to scale up to higher storage as needs arise, and taxpayers can gain budget efficiencies in the process. With many government file servers currently underused, consolidating federal data centers and migrating applications to the cloud would save money while also improving service delivery.

Electronic Communication Tools in Congressional Offices

Congressional offices have been slow to embrace many forms of technology innovation. The legislature is still bound by rules that place limits on various types of digital communications. Only recently have some of these rules been changed, allowing the use of video and social networking sites, such as YouTube and Facebook. Employment of phone conferencing applications such as Skype have not been allowed in House offices.[32]

Members can use this service on mobile or WiFi Internet networks but not on desktop computers. Leaders in Congress have labeled Skype "unauthorized" because of fears that the service would be vulnerable to security threats. If the phone conferencing feature is a security risk on desktops, why is its use allowed on mobile devices? Inconsistencies such as these make little sense in government computing policy.

Judicial Computing Usage

Federal courts lag on many indicators of information technology compared with the rest of the public sector. They are less likely to make use of online services and interactive features, and their websites are less likely to provide mechanisms for visitor feedback or to allow visitors to personalize the site to specific interests. Of the sixty-one federal sites rated, thirteen of the bottom sixteen performers were federal court sites, including the Supreme Court and appellate courts from around the country.[33]

There is far less cloud usage in the judicial than in the executive or legislative branches of government. Court officials need to consider ways to modernize judicial computing usage in general but cloud computing in particular. Economies and efficiencies would be gained by migrating some court applications to the cloud. As noted earlier, some agencies have found cost savings of 25 to 50 percent from transferring e-mail services to cloud platforms.

Government-Wide Contracting and Certification Processes

Many federal departments order software and hardware on an agency-by-agency basis. This makes it difficult for the federal government to take advantage of its huge purchasing power to gain economies of scale. Government-wide contracts would allow public officials to drive harder bargains and become more efficient in the use of federal resources.

Federal agencies lack a uniform certification process. Right now, each agency is responsible for certifying its own applications.[34] Approval by one department gives no guarantee that its standards will meet the certification standards of another department. This creates unnecessary inconsistency and delays in the procurement process as vendors go through multiple checkups for the same product.

Some of these problems could be addressed by the creation of a joint authorization board with the power to review management services and certify particular products for use across the government. The new Federal Risk and Authorization Management Program known as FedRAMP represents a big advance toward this type of government-wide approach to certifying and accrediting particular applications. It should be used to expedite technology innovation in government.

International Rules and Norms on Cloud Computing

Some nations are becoming increasingly protective in their treatment of cloud computing. In the name of national security, they want to store data only within their own national borders and take the view that the only system they trust is one operated within their own jurisdictions.

American jurisdictions have put restrictions in place. Cloud agreements negotiated in Los Angeles over the past two years have included a promise to store government cloud data within the continental United States. Local government officials justify this stance on security grounds, but it limits the efficiency of cloud computing systems.

Europeans have a data protection directive that precludes data transfers from the European Union to countries with privacy and confidentiality rules it views as inadequate.[35] Since this includes the United States, this EU rule limits the ability of American cloud providers to build cross-national networks and take advantage of the economies of scale that are a strong feature of cloud computing.

Research has demonstrated that European privacy rules affect consumer behavior in the area of online advertising. Work by Avi Goldfarb and Catherine Tucker shows that rules restricting the ability of companies to collect consumer information in order to target advertising campaigns were linked to a drop in ad effectiveness.[36]

To facilitate innovation and get the greatest economies of scale from new technology, rules on cross-border transactions via the cloud will need to be clarified and made more compatible where possible. We need international agreements harmonizing national rules so that the current Tower of Babel does not undermine fiscal efficiencies or use of cloud platforms. An agreement between nations on basic privacy, security, and access control rules would pave the way for cloud networks and data sharing on an international scale.

Portability of Information across Platforms

Mechanisms for data exchange that encourage portability from one computer platform to another are also lacking. These include desktop-to-cloud, cloud-to-desktop, and cloud-to-cloud transfer. We especially need portability when people or organizations shift cloud platforms. To avoid vendor lock-in that precludes data exchange, recognized procedures for file structure and metadata must be instituted. The creation of cloud import and export tools would expedite the portability of information across platforms.[37]

Transparency of Cloud Performance

Taxpayers are understandably nervous about security, privacy, and performance with cloud applications. They worry whether confidential information stored by federal agencies on everything from taxes and health care to retirement pensions will be protected. Many do not understand cloud computing and are concerned that the metaphor of a wispy and ever-moving cloud will not protect their private information.

In this situation, public officials need to reassure skeptical citizens regarding the reliability of cloud storage. Effective service-level agreements should make clear what the performance expectations are and whether providers meet those expectations. Data on uptime, downtime, recover time, archiving, and maintenance schedules would help build public trust in the cloud, as would information on what providers are doing to safeguard security standards.

For example, when the General Services Administration moved its portal site USA.gov to the cloud, officials found that it "was able to reduce site upgrade time from nine months to one day; monthly downtime improved from two hours to 99.9% availability; and GSA realized savings of $1.7M in hosting services."[38] Indeed, most commercial providers commit to 99.9 percent availability in their service-level agreements. Having this type of information about computing performance reassures taxpayers and helps legislators understand the advantages of cloud computing.

CHAPTER FIVE

HIGH-SPEED BROADBAND

IN THE AMERICAN RECOVERY AND Reinvestment Act of 2009, the U.S. Congress charged the Federal Communications Commission with developing a national broadband policy by February 17, 2010. Legislators asked the commission to outline policies that would be efficient, effective, and affordable and would advance the public interest in "consumer welfare, civic participation, public safety and homeland security, community development, health care delivery, energy independence and efficiency, education, worker training, private sector investment, entrepreneurial activity, job creation and economic growth, and other national purposes."[1]

Central to this request was the idea that digital infrastructure is vital to long-term economic, social, and civic development. In an age of technology, people need to stay connected in order to innovate and take advantage of new digital applications.

Countries vary in size, population density, industry structure, regulatory environment, demographic makeup, and cultural patterns. It is important to look internationally to see valuable lessons that can be learned from other countries in terms of new ideas and possible benefits to be gained from high-speed broadband. How are other countries using broadband technology to advance social and political innovation? What broadband speeds are they aiming for in their national plans? How are they paying for necessary broadband investments? What new applications become available to them at various broadband speeds? How valuable do they see broadband for the economy, social connections, civic engagement, and public sector service delivery?[2]

South Korea has the most ambitious national goal for future broadband traffic. It is seeking to raise broadband speed to 1 gigabit per second. Australia and Finland are aiming for 100 megabits per second (Mbps), while Germany has a stated target of 50 Mbps by 2014. These countries are pushing for high speeds because they see them as necessary for new applications in health care, education, energy, and civic engagement.

The question for most nations is how to improve access, open networks, and pay for faster service. The Federal Communications Commission estimates that it will require $350 billion to provide universal broadband coverage in the United States at 100 Mbps, yet the public investment authorized as part of the economic stimulus package is only $7.2 billion.

Christine Zhen-Wei Qiang, reviewing the role of broadband infrastructure investments in the stimulus packages of leading countries, suggests that many locales have identified broadband as a key to future development.[3] She finds that the United States is investing the largest sum of money in infrastructure ($7.2 billion), followed by Japan ($3.7 billion), Australia ($3 billion), Canada ($150 million), Finland ($130 million), and Spain ($90 million) (see table 5-1). But even the U.S. investment pales in comparison with what is needed. Encouraging greater private sector investment is key to realizing the long-term benefits of broadband.

Faster speeds are vital to taking advantage of new digital tools such as geographic information system mapping, telemedicine, virtual reality, online games, supercomputing, video on demand, and video conferencing. New developments in health information technology and mobile health, such as e-mailing X-rays and other medical tests, require high-speed broadband. Distance

TABLE 5-1. Broadband Plans, by Country

Country	Time frame	Broadband goals
Australia	Eight years from 2010	Deliver broadband at speed of 100 Mbps to 90 percent of homes, schools, and businesses
Canada	Four years from 2009	Extend broadband coverage to all currently underserved communities
Finland	Seven years from 2009	Provide broadband to every household, with minimum download speeds of 1 Mbps by 2010 and 100 Mbps by 2016
France	Five years from 2008	Provide universal access to broadband at affordable price by the end of 2013
Germany	Ten years from 2009	Provide broadband access at 50 Mbps to 75 percent of households by 2014 and rest by 2019
Ireland	Two years from 2009	Provide broadband to all with minimum 1.2 Mbps
Japan	Two years from 2009	Extend broadband to all rural areas
Portugal	Two years from 2009	Extend broadband to 50 percent of residences by 2010 and others by 2011
Singapore	Five years from 2009	Universal connection to Next Generation broadband by 2014
South Korea	Five years from 2009	Upgrade broadband to 1 Gbps
Spain	Four years from 2009	Extend broadband to rural areas

Source: Christine Zhen-Wei Qiang, "Broadband Infrastructure in Stimulus Packages: Relevance for Developing Countries," World Bank, 2009.

learning, civic engagement, and smart energy grids require sufficient bandwidth.

The challenge for the United States is to create the proper mix of speed and access. Some question the wisdom of aiming for fiber-optic lines with 100 Mbps of speed when one-third of Americans lack home broadband access.[4] More people would benefit from the digital revolution if the United States placed priority on improving adoption rates. To balance the competing goals of access and speed, priority also should be given to high-speed networks for hospitals, schools, and libraries. Putting institutional anchors in place in each community with fast broadband would allow many people (including those without home access) to take advantage of new digital applications while maximizing scarce financial resources.

USING BROADBAND TO JUMP-START NATIONAL ECONOMIES

Many nations around the world are investing in broadband infrastructure as a way to jump-start economies weakened by the recent financial collapse. The decline in stock market valuations, the rise in unemployment, and the reduction in overall economic growth highlight the need to target financial resources and develop national priorities. In conditions of economic scarcity, countries no longer have the luxury of being passive and reactive. They must be proactive and forward-looking and think clearly about how to create the basis for sustainable economic recoveries.

Taylor Reynolds has analyzed the role of communication infrastructure investment in economic recoveries among member nations of the Organization for Economic Cooperation and Development (OECD) and finds that many view broadband development as crucial to their economic stimulus packages.[5] He demonstrates that there is a strong connection between telecommunication investment and economic growth, especially following recessions. These kinds of investments help countries create jobs and lay the groundwork for long-term economic development.

Key Policy Questions

The government of Canada recently became the first Western nation to say it would regulate Internet network management. Its Canadian Radio-television and Telecommunications Commission voted to require Internet service providers to tell consumers how they manage traffic and inform them of the impact of their techniques on customer service. According to commission chairman Konrad von Finckenstein, "Canada is the first country to develop and implement a comprehensive approach to Internet traffic management practices. . . . The centerpiece of our approach is a framework of analysis that will be employed to determine whether economic and technical practices are acceptable."[6]

Finland also made history by declaring that its citizens have a "legal right" to broadband. It passed a law requiring telecommunications providers to "provide all Finnish residents with broadband lines that can run at speeds of at least 1 megabit per second" by July 2010.[7] Because 96 percent of its people already have access to broadband, the legislation affects only the 4 percent who currently are unconnected. The nation's communications minister announced plans to bring 100 Mbps Internet speed to all residents by 2016.

France has not declared broadband access a legal right, but its courts have defined it as a basic human right. In a 2009 Constitutional Council action, officials ruled that people need broadband in order to participate in civic affairs and that the country should ensure that all its people have access. Around the world, this and other actions demonstrate that countries are pursuing broadband development according to a variety of models. Nations vary enormously in how they value broadband speeds, how they fund infrastructure development, and how they handle underserved communities.

Actual network speeds vary considerably across countries. According to a study conducted at SpeedMatters.org and reported by the Communications Workers of America, South Korea had the highest average download speed of 20.4 Mbps, followed by Japan (15.8 Mbps), Sweden (12.8 Mbps), the Netherlands (11.0 Mbps), Germany (around 9 Mbps), and the United States (5.1 Mbps).[8]

What broadband speed a country deems necessary depends very much on national aspirations and views regarding the applications that are envisioned over the short and long terms. South Korea has developed the most ambitious plan in terms of network speed, with a national goal of 1 Gbps. Australia and Finland are aiming for 100 Mbps, and Germany has a stated target of 50 Mbps by 2014.

According to the OECD, the top bandwidth required for various new applications runs as high as 18.0 Mbps for high-definition television, 14.0 Mbps for online games, 13.5 Mbps for video on demand and Internet Protocol TV, 13.4 Mbps for video conferencing, 9.0 for access to virtual worlds, 4.0 for web browsing, 1.5 for audio streaming, and 0.5 for voice calls.[9] As these applications gain popularity, the demand for high-speed broadband will grow.

The California Broadband Task Force predicts that telemedicine, education distance learning, and digital medicine will require speeds between 10 and 100 Mbps.[10] It furthermore concludes that high-definition telemedicine, virtual reality, supercomputing, and advanced research applications will require broadband speeds of more than 100 Mbps.

A Nielsen study finds that the growth of broadband usage in New Zealand has stimulated blogging, social networking, file sharing, and using multimedia applications.[11] One interesting finding of this research is that video uploading has reached 43 percent of the online population, blog creation has increased to 28 percent, online profile creation to 50 percent, and social networking to 48 percent of the online population.

According to this analysis, 62 percent of people fourteen years or older have used the Internet for browsing other people's online profiles in the past twelve months, while 45 percent have created profiles and 48 percent have joined "friend finder" sites. Half of all users are now reading blogs, and 17 percent are creating them. Nearly 80 percent are sharing photos, 61 percent are uploading photos, 59 percent share links, 39 percent share videos, 31 percent upload music, and 24 percent upload video content.

One of the reasons why South Korea is moving toward 1 Gbps networks is that it is the online game-playing capital of the world. Seoul has 26,000 "bangs," or online game rooms.[12] The country has three television cable channels devoted exclusively to games.

TABLE 5-2. Digital Music Sales, by Country

Units as indicated

	Total sales (in millions of U.S. dollars)	Share of total music sales (percent)
United States	1,849	16
Japan	778	15
United Kingdom	201	6
South Korea	151	61
France	128	8
Germany	117	6
Canada	51	7
Italy	41	7
Australia	37	6

Source: Organization for Economic Cooperation and Development, "Information Technology Outlook," Paris, 2008.

Young people gather at game rooms, at home, or in front of television sets to play large, interactive games with others. Indeed, Korea recognizes these social gathering places as "third places," meaning locations other than work or home where people connect. It is estimated that more than 5 million people play Starcraft, one of the popular online games in Korea. Other popular games include Lineage, Ultimate Online, Kingdom under Fire, and EverQuest.

Demand for online music downloads is growing in many countries. The United States already has $1,849 million in sales (16 percent of all music sales), compared with $778 million for Japan, $201 million for the United Kingdom, and $151 million for South Korea (see table 5-2).[13]

Online film sales are expected to grow substantially in the near future. Revenue from online film sales in the United States was $22.9 million in 2006 but is expected to grow to $1.975 billion by 2010. Similar trends are anticipated for the European Union. Its sales totaled $24.6 million in 2006 but are expected to grow to $1.356 billion in 2010.[14]

Online video gaming is taking off among the youth of many nations. From a base of less than $1 billion in sales in 2002, sales are expected to grow to more than $15 billion.[15] Digital games require fast broadband because they use virtual reality features, 3-D graphics, and multiplayer platforms. For example, the game Anarchy Online features thousands of players roaming through 480 square meters of territory and battling adversaries along the way. It is estimated that more than 720,000 people have played this game online.

Based on international experience, it is reasonable to conclude that demand for high speeds will rise dramatically as new applications come online. An analysis of other countries suggests that online music, video, and games are very popular, especially among young people, and each of these applications requires faster speeds than what the United States has right now.

Some countries view broadband as a public good to be financed by the central government alone or in conjunction with phone companies. Public officials in these places have placed a high priority on high-speed broadband and provided direct resources or indirect incentives for those investments. The Japanese Ministry of Internal Affairs and Communications and the Nippon Telegraph and Telephone company, for example, worked together to develop fast broadband.

A similar phenomenon happened in South Korea through KT, a company created by the merger of Korea Telecom and KTF, a wireless carrier. Kepco, Korea's utility company, developed a fiber-optic cable network many years ago for its own use and now leases the unused 90 percent of this network to service providers. Korea now has the highest bandwidth of any nation. Between KT and Kepco, consumers can purchase high-speed broadband for $25 a month. It has been estimated that with cheap and speedy broadband, nearly all Koreans have access to fast broadband service.[16]

Between 1995 and 2005, the Korean government invested $900 million in broadband, stimulating $32.6 billion in private technology investment.[17] Public officials aggressively pushed high-speed applications in government and business. This sparked substantial increases in consumer demand. In exchange for the government dollars, private companies were required to connect public institutions.

Of course, Korea has benefited from a set of favorable conditions not present in other countries. Its high population density and homogeneous population make providing Internet service economically feasible. Whereas population density in the United States is 31 people per square kilometer, in Korea it is 476, and in Japan, 337.[18] More than 70 percent of Koreans reside in half a dozen cities. Its broadband backbone contains 13,670 miles of fiber-optic lines, less than Verizon's coverage in the state of West Virginia alone.[19] Some rental rates in Korea are based on bandwidth capacity, and enterprising landlords with fast pipes pitch Internet speed to attract renters.

Most other nations have opted for a broadband system based mainly on private investment. The idea is that governments should not finance broadband structure, nor should it compel commercial carriers to undertake broadband investment. Rather, they should wait until favorable market conditions induce the private sector to make such investments. This allows companies to make relevant decisions based on their perceptions regarding return on investment.

Historically, the United States has relied on private companies for the bulk of broadband infrastructure. AT&T, for example, invested more than $44 billion in the past two years building a broadband and wireless network. Verizon has spent nearly $43 billion, Comcast has invested $14 billion, Sprint $10 billion, and T-Mobile $8.5 billion. These investments have created networks through which 63 percent of Americans have broadband access at home.[20]

A recent report by the U.S. Federal Communications Commission estimates that it would cost $350 billion to provide 100

Mbps universal broadband coverage in the United States. The federal government invested only $7.2 billion from the economic stimulus package, and it is apparent that the vast bulk of American infrastructure funding will come from the private sector.[21]

A serious challenge for most countries has been bringing service to rural areas or other underserved geographic places. The cost of wiring the "last mile" is high because of low concentrations of people and limited demand for service. It is here that the public sector in most countries has played a significant role. According to Christine Qiang,

> For the goal of universal coverage of broadband services, countries are spending larger public funding for rolling out high-speed networks to areas that are underserved or unserved by commercial internet service providers. The Finnish government plans to foot one-third of such costs. Others are contracting commercial providers to build the network with service obligations through a competitive bidding process (for example, France, Ireland, Japan and Singapore). The [European Union] and the US are adding resources to existing rural development or universal service.[22]

Impact of Broadband on Economic Development

The influence of broadband access on economic development supports the complementary growth of advanced digital infrastructure. Indeed, examples from other countries show how fast broadband spurs development and makes possible new businesses, products, and services.

A Strategic Networks Group study of broadband investment in a fiber-optic network in the South Dundas township of Ontario, Canada, calculates that an investment of $1.3 million led over several years to a "$25.22 million increase in GDP for Dundas County and $7.87 million increase for the Province of Ontario" and the creation of 207 jobs.[23] Overall, researchers using a survey of businesses and organizations in the county observed an increase of $3.5 million in provincial tax revenues and $4.5 million in federal tax revenues that were directly attributable to the new fiber-optic lines. Fifty-four percent of the area's businesses that had access to the fiber-optic network reported job growth, compared with 27 percent of businesses that had dial-up Internet access and 5 percent of those with no Internet access.

Raul Katz and his colleagues examine the impact of broadband on jobs and GDP in Germany.[24] Overall, they estimate that the next ten years will see gains of 968,000 jobs and 170.9 billion euros added to the economy. This amounts to 60 percent growth in GDP during the period 2010–20. Some of this growth arises in the form of network construction, while the remainder is generated by rising broadband penetration and subsequent innovation in business activity.

A study of 120 nations undertaken by Qiang estimates that between 1980 and 2006, each 10 percentage point increase in broadband penetration added 1.3 percent to the GDP of high-income countries and 1.21 percent in low- to middle-income nations.[25] This suggests that growth comes not just in direct form, as estimated by other authors, but also from the new applications for businesses and consumers generated by broadband.

Many studies note the difficulties in estimating the economic ramifications of broadband development.[26] Peter Collins, David Day, and Chris Williams, of the Australian Department of Communications Information Technology, review the literature on broadband and argue that studies need to incorporate "counterfactuals" into the analysis.[27] It is not sufficient to draw a direct link between investment and job growth. Rather, researchers must examine how much growth would have occurred in the alternative scenario of no broadband development and incorporate that into their conclusions. According to Collins, Day, and Williams, considering various possibilities provides a more useful basis for evaluating broadband.

Impact on Social and Cultural Connections

High-speed broadband facilitates social and cultural connectedness. D. H. Shin describes the way in which broadband has facilitated the emergence of "digital cities" in Korea and the United States.[28] Using a comparative case study of public networks, the author notes that broadband brings people together into knowledge networks and shows how these social connections have helped people in economically depressed areas overcome spatial isolation.

Katia Passerini and Dezhi Wu identify a similar phenomenon in their analysis of "intelligent communities."[29] They present best practices in terms of social capital development. In their conception, "megacommunities" consisting of diverse, multilateral stakeholders use broadband technology to work for solutions to a range of health, environmental, and social problems. These enterprises form the backbone of the new knowledge economy and use information and communication technologies to stimulate higher-level economic growth.

Paschal Preston and Anthony Cawley forecast a "virtuous circle" arising from efforts by the European Commission to develop a knowledge-based economy.[30] The European Commission's plan for an iNetWorked Society based on high-speed broadband emphasizes demand-side applications over supply-side infrastructure. With fast networks, government planners are stimulating new uses in business, work, and government.

Stefan Agamanolis draws on the European experience to describe how digital technologies enhance community, offer users the chance to encounter people outside their common circle, and allow people to build relationships in new sorts of ways.[31] The Human Connectedness research group, active from 2000 to 2005 at Media Lab Europe, developed new applications designed to overcome spatial distance in human relationships. Among the prototypes constructed were virtual reality devices, adaptive speech interfaces that combine the spoken word with nonverbal gestures, biofeedback mechanisms, and audio bulletin boards that allow people to share short audio messages with other people.

Valerie D'Costa and Tim Kelly find that broadband provides a new platform for economic, social, and cultural development in Asia.[32] Reviewing data on broadband subscribers and penetration rates, the authors argue that fast connections enable people to connect in new ways and build digital communities.

Impact on Civic Engagement

Fast broadband promotes civic engagement and new ways to follow politics and government. A number of public agencies around the world have developed interactive mapping software that allows citizens to chart data patterns in their neighborhood and innovative uses of video or multimedia to engage people in public debates.

Geographic information systems (GIS) for purposes of civic engagement are becoming increasingly widespread. These are interactive sites that allow people to map a range of social, economic, political, demographic, and policy features onto local, state, national, or international jurisdictions. For example, a number of cities have mapping capacities on government websites that enable site visitors to see crime or other rates broken down by individual blocks. This allows users to chart crime statistics, student achievement, or other data trends along social, economic, or political dimensions.

Renate Steinmann, Alenka Krek, and Thomas Blaschke look at public participatory GIS, which focus on ways to get citizens involved in civic decisions.[33] They evaluate twelve GIS applications in the United States and Europe on interactivity, usability, and visualization. Among the projects analyzed are urban design visualization, resource management mapping, river basin analysis, and landscape planning. For example, the village of Bradford, England, offers online maps that allow people to zoom and select specific features for study. Salford University employs an open-space platform with 3-D capacity that allows users to walk through a virtual city while submitting design suggestions to city planners. The Freising district in Germany enables visitors to control development options on interactive city maps.

Oscar Cardenas-Hernandez and Luis Martinez Rivera have studied the use of remote sensing and satellite imagery in natural resource management.[34] Using the example of environmental degradation in western Mexico, they demonstrate how interactive GIS maps allow citizens to participate in the sustainable management of rural communities. By interweaving information from conservation, agricultural, and irrigation zones, these maps help people understand the trade-offs across policy domains.

Public officials increasingly are using online communications to keep in touch with constituents. In 2008 G. S. Hanssen conducted a national survey of municipal politicians and mayors in Norway to explore how local politicians use digital communications to engage citizens and industry stakeholders in policymaking.[35] His results indicate that e-mail is the most important communications channel between local politicians and citizens. He also finds that mayors use e-mail in work-related communications more than other public officials.

Tapio Hayhtio and Jarmo Rinnel argue that the Internet has become increasingly important to political participation and mobilization.[36] Using the example of a Finnish protest against gossip journalism, the authors suggest that high-speed networks allow dissatisfied elements within society to organize, identify supporters, and build coalitions. The campaign relied on the Internet to protest the publication of unmasked pictures of the musician Lordi following the Eurovision Song Contest. Lordi was a member of a Finnish heavy metal band that wore monster masks on stage. Despite a request not to publish images of the group without their performance masks, several tabloid outlets printed pictures of members of the band out of costume. This riled group supporters, who used social networking sites and digital communications to organize a protest against the tabloid publications.

Hernando Rojas and Eulalia Puig-i-Abril examine the impact of digital communication technologies on political mobilization and civic participation.[37] Using data from a random public opinion sample of Colombia's adult urban population, these authors document how broadband Internet and mobile phones aid "expressive participation" in online protests. They undertook a survey of online information usage and found a relationship between digital information acquisition and political engagement. Those who sought information from the Internet were more politically active and expressive than those who did not. The authors conclude that in developing societies with high levels of political, economic, and social conflict, digital communications represent a valuable pathway for democratic political engagement.

Heekyung Kim, Jae Moon, and Shinkyu Yang argue that the Internet contributed to the political victory of Korean president Roh Moo-hyun in 2002.[38] The Internet allowed the candidate to identify and mobilize supporters and the people to gain news and videos from unfiltered sources, such as the citizen journalism site OhmyNews. Because of this, Roh was known as Korea's first Internet president.

Impact on Public Service Delivery

High-speed broadband is vital to the emergence of online services in the public sector. Some countries have developed "smart cards" that allow citizens to complete official transactions online quickly and efficiently. Taiwan has been a leader in this area. In the past year, more than 22.8 million Taiwanese people used broadband online services to file taxes, find land information, use health services, and track diseases, among other things.

Taiwan and Singapore have done the same thing through mobile devices. Taiwanese agencies offer wireless navigational systems, tour guide information, friend finders, and health care services through smartphones. In Singapore, public agencies allow citizens to receive personalized short text-messaged alerts, renew passports, and pay road taxes. Singapore also has an online civil defense force that encourages people to use the Internet to report bomb threats, get information on evacuation routes, recognize signs of a chemical or biological attack, or locate medical facilities.

Peter Trkman and Tomaz Turk present a conceptual model for the development of broadband and e-government.[39] Looking at the experience of various countries around the world, they suggest that there is a strong relationship between broadband diffusion and the development of e-government and e-commerce. Countries that have fast broadband typically make greater progress in building their public and private sectors.

Estonia has a direct democracy portal called "Today I Decide" that facilitates participation in government activities. The website publishes draft laws online and invites public comment on proposed regulations. It also provides a citizen-initiated process through which Estonians can propose specific laws or regulations for review by government officials. Fully one-quarter of the suggestions submitted through this portal have qualified for serious consideration by government ministries, and 3 percent have turned into formal legislative proposals for deliberation by elected officials.

Unusual among countries, Estonia has made extensive use of Internet-based voting. According to Michael Alvarez, Thad Hall, and Alexander Trechsel, citizens have been offered the option of using digital technology in two national elections.[40] Voters used their national identity card with digital signature to cast ballots through secure smart cards. After undertaking quantitative surveys, the authors find that e-voting enhanced electoral participation by young people and those who trusted online technology.

The Sheffield City Council in the United Kingdom has set up a web-based voting system by which citizens can cast ballots for elected officials. People were offered the choice of voting through

TABLE 5-3. Percentage of International Government Websites with Online Services

Number of services	2001	2002	2003	2004	2005	2006	2007	2008
None	92	88	84	79	81	71	72	50
One	5	7	9	11	8	14	11	19
Two	1	2	3	4	3	5	4	9
Three or more	2	3	4	6	8	10	13	22

Source: Darrell West, "Improving Technology Utilization in Electronic Government around the World, 2008," Brookings, 2008.

paper ballots at a polling place, electronic voting on a personal computer, by telephone, or through a mobile text-messaging service. More than 2 million people participated in the election, the highest number in recent years. Turnout through electronic means was especially high in precincts with a large number of students. Overall, 40 percent of students chose to cast their votes through electronic means. Postelection surveys have found that 95 percent of voters were satisfied with the system and willing to vote electronically in the future.[41]

In 2008 I undertook an international study exploring the number and type of online government services offered over the past decade.[42] Government use of online service delivery increased dramatically over the period (see table 5-3). Of the government websites studied, 50 percent had services in 2008 that are fully executable online, up from 28 percent in 2007, 29 percent in 2006, 19 percent in 2005, 21 percent in 2004, 16 percent in 2003, and 12 percent in 2002. In 2008, 19 percent offered one service, 9 percent had two services, and 22 percent had three or more services.

However, not all geographic areas have benefited from this progress. There are major variations around the world, and it is clear that areas with fast broadband were more likely to have online services. North America (including the United States, Canada, and Mexico) offered the highest percentage of online services. Eighty-eight percent had fully executable, online services (see table 5-4). This was followed by the Pacific Ocean Islands

TABLE 5-4. Percentage of Government Sites with Online Services, by Region

Region	2001	2002	2003	2004	2005	2006	2007	2008
North America	28	41	45	53	56	71	62	88
Pacific Ocean Islands	19	14	17	43	24	48	28	66
Asia	12	26	26	30	38	42	36	49
Middle East	10	15	24	19	13	31	29	50
Western Europe	9	10	17	29	20	34	34	59
Eastern Europe	n.a.	2	6	8	4	12	11	32
Central America	4	4	9	17	15	11	22	63
South America	3	7	14	10	19	30	46	75
Russia and Central Asia	2	1	1	2	3	11	10	10
Africa	2	2	5	8	7	9	9	30

Source: Darrell West, "Improving Technology Utilization in Electronic Government around the World, 2008," Brookings, 2008.

(66 percent), Western Europe (59 percent), the Middle East (50 percent), Asia (49 percent), Eastern Europe (32 percent), Africa (30 percent), and Russia and the former Soviet republics (10 percent). In places where there is limited broadband, it is harder for governments to innovate and bring citizens the advantages of the digital revolution.

A number of innovative services and applications based on high-speed connections have been developed on government websites. Antigua and Barbuda's Department of Tourism has online newsgroups where people planning trips can have online discussions. Ecuador's Ministry of Defense site has streaming radio with options for news or different genres of music. Fiji's computerized human resources information system matches people to jobs. The Egyptian tourism website offers a hieroglyphic translator. Peru's tourism commission website offers users a desktop calendar download that helps them organize a trip to Peru.

Inside the Canadian health website, there are sections with an interactive menu providing help to visitors concerning health care. The Canadian portal has audio feeds that allow the user to change the voice and speed of selected readings. The New Zealand conservation site has an option that allows visitors to order reports and information according to region. The Austrian agriculture site allows users to calculate their ecological footprint based on various activities.

CONCLUSION

The experience of other countries demonstrates that high-speed broadband enhances economic development, social connections, civic engagement, and online government. Broadband is no longer just a technology issue but is stimulating the creation of new applications in areas such as health care, education, energy, and entertainment.

High-speed broadband facilitates the adoption of remote wireless health monitors, GIS mapping, social media, distance learning, smart energy grids, file sharing, and video conferencing. A number of public agencies around the world have developed interactive software that allows citizens to map data for their neighborhood. Video and multimedia have been used to foster engagement in civic debates. These applications require faster speeds than those currently available in many places.

The United States needs to place a priority on boosting individual adoption and on raising broadband speeds for institutional anchors in each community—schools, hospitals, and libraries. Expanding the speed and the reach of Internet connections to publics places will allow people without home access to take advantage of new digital applications while maximizing scarce financial resources.

PART THREE

POLICY CONSIDERATIONS

CHAPTER SIX

INNOVATING IN THE PUBLIC AND PRIVATE SECTORS

IT IS OFTEN ARGUED THAT the private sector is more entrepreneurial and innovative than the public sector. Commercial enterprises have great incentives to innovate because of market pressures and the need to remain competitive. As a result, they develop organizational structures that place a high priority on incorporating new technologies into their operations as a way to boost efficiency and productivity.

The public sector, in contrast, encounters challenges in regard to innovation and entrepreneurship. Government agencies do not face the same kind of market pressures as private companies do. They do not have customers in the traditional sense, and they are not required to show a profit on their revenues. Most public departments have multiple constituencies, such as people who use their services, voters, taxpayers, legislators, administrators, the media, advocacy organizations, and nonprofit organizations.

Yet despite the rhetoric about public and private sector differences, few researchers have collected data comparing technology innovation in the two sectors. Systematic data that evaluate innovation in business and government to see which sector leads and on what dimensions would suggest broader lessons about the factors that facilitate technology innovation and how to encourage the entrepreneurship model in the future.[1]

METHODOLOGY

To study technology innovation, I examined the websites of 68 leading American corporations, 1,476 state government websites, and 61 federal government sites. For the business analysis, I drew a stratified sample of companies and analyzed how their websites handled a variety of digital features. I included companies of varying sizes and types to get a representative view of the private sector's use of technology. For the public sector, I analyzed leading federal government sites and an average of thirty websites within each state government. Sites included those of court offices, legislatures, elected officials, major departments, and state and federal agencies serving crucial functions of government, such as health, human services, taxation, education, corrections, economic development, administration, natural resources, transportation, elections, and agriculture.

I tracked eighteen features that corporations, state governments, and federal agencies offer on their websites: publications, databases, audio clips, video clips, foreign language access, the absence of ads, the absence of user and premium fees, disability access, privacy policies, security policies, functions allowing digital signatures on transactions, an option to pay by credit card, e-mail contact information, areas on which to post comments, an e-mail update option, a function allowing for personalization of the website, and PDA (personal digital assistant) or handheld device accessibility. Four points were assigned for each of these features, up to a maximum of 72 points.[2]

Each site qualified for additional points based on the number of online services executable on that site (0 for no services, 1 point for one service, 2 points for two services, 3 points for three services, and so on up to a maximum of 28 points for twenty-eight or more services). After adding the features and online services, the technology index ran along a scale from 0 (having none of these features and no online services) to 100 (having all eighteen features plus at least twenty-eight online services).

I also undertook case studies of successful innovation among leading corporations in order to understand what worked and how common obstacles to technology innovation were overcome. Interviews with industry leaders helped identify keys to innovation and successful strategies for developing and maintaining high-quality websites.

COMPARISON OF PUBLIC AND PRIVATE SECTOR WEBSITES

Many government websites lag the private sector in offering multimedia, interactivity, and personalization. Internet users like interactivity and the capacity to access information, order electronic services, and offer feedback on website features. These are among the most useful Internet features because they allow users to tailor information to their own needs. Indeed, that is the central virtue of the technology revolution: it breaks down hierarchies and gives consumers the power to manage their own information.

However, the private sector does not outperform the public sphere in every area. In a few domains the public sector surpasses its private counterparts. For example, government sites are better at providing access to the physically disabled. Computer software allows the visually impaired to have website content read to them in audio form. But sites must be configured properly for this software to function. In this analysis of government and business, I find that public sector agencies are more effective than commercial enterprises at providing disability access.

The same public sector advantage exists in regard to privacy policies. Public sector websites offer stronger consumer protections than commercial sites do. Government agencies generally have clearer policies prohibiting the use of "cookies" that identify visitors. They also have superior policies prohibiting the sharing of personal information with third parties.

TABLE 6-1. Ratings of Private and Public Sector Websites[a]

Top corporations	Top state governments	Top federal agencies
Wells Fargo 92 points	Delaware 83.7 points	USA.gov 92 points
Home Depot 84	Georgia 78.3	Department of Agriculture 79
Walgreens 84	Florida 77.9	General Services Administration 77
AT&T 82	California 70.9	U.S. Postal Service 76
American Express 81	Massachusetts 69.5	Internal Revenue Service 73
Federal Express 81	Maine 67.7	Department of Education 72
CVS Caremark 80	Kentucky 67.3	Small Business Administration 71
Symantec 78	Alabama 66.4	Library of Congress 70
Google 77	Indiana 65.0	Department of Treasury 69
Microsoft 77	Tennessee 64.3	Federal Reserve 69

Source: Darrell West and Jenny Lu, "Comparing Technology Innovation in the Private and Public Sectors," Brookings, 2009.

a. Numbers given are website ratings on a scale of 0 to 100.

Overall, I found that corporations scored the highest (see table 6-1). The business sites examined earned an average of 65 of 100 points, the state government sites 54 points, and the federal sites, 51 points.

Online Information

For the study of public and private sector websites, I analyzed the availability of publications, databases, and audio and video clips. In general, access to publications and databases is excellent across the sectors. Nearly all private and public sector sites studied offer publications, and most have databases.

The private sector outpaces the public sector in providing audio and video clips. Whereas 98 percent of corporate sites have audio clips, only 40 percent of state government and 70 percent of federal sites do. The trend is similar in regard to video clips. Eighty-two percent of corporations have video clips, compared with 48 percent of state governments and 72 percent of federal government sites.

The corporations studied often had webcasts of investor conference calls. On their jobs pages, a number featured videos of current employees talking about their jobs. In the public arena, audios and videos typically featured politicians giving speeches or webcasts of government meetings, such as those of the state legislature or congressional committees.

Electronic Services

Fully executable online service delivery benefits both government and its constituents. In the long run, this type of delivery offers the potential for lower cost of service delivery and makes government services more widely accessible to members of the general public, who no longer need to visit, write, or call an agency to execute a specific service.

Of the websites examined, all corporate sites featured online services, compared with 98 percent of federal sites and 88 percent of state government websites. Nearly all the company sites had three or more online services, whereas only 66 percent of state government sites featured that many services; and 88 percent of federal websites had at least three electronic services. Business sites had an average of fourteen electronic services. This was lower than the average of twenty-four services for state government sites but higher than the average of ten online services for federal agencies.

Novel Services or Features

Commercial sites contained a number of interesting features. In addition to common services for ordering merchandise, a number had web links to sites where people could check rumors, submit innovative ideas, participate in politics, and file reports about suspected illegal or unethical behavior. For example, Zimmer Holdings had an online feature, Zimmer Compliance Hotline, that allowed people to report violations of applicable laws, of Zimmer's Code of Business Conduct, or of federal health care program requirements. U.S. Steel offered an EthicsPoint, where visitors could report suspected illegal or unethical conduct. Several companies allowed people to submit online ideas for inventions or business innovation.

Some companies use their website to debunk rumors and myths about the company. For example, Coca-Cola has a Facts and Myths section that disputes stories alleging that its product contains material unsuitable for vegetarians or Muslims. A Rumor Response section on the Starbucks site corrects false stories about the relationship between the company and the U.S. military.

The financial services companies surveyed had a number of online services, such as online banking, bill pay, and brokerage. Customers could open a checking account, apply for a credit card, loan, or line of credit; get insurance quotes; use calculators for home equity amortization, debt consolidation, and home improvement; and order foreign currency or travelers checks. Wells Fargo has a "virtual safe" called vSafe for the online storage of important personal documents.

Several companies participate in the eTree program, whereby for each shareholder who signs up for electronic delivery of proxy materials, the company agrees to plant a tree. According to their websites, participating companies include Coca-Cola, Verizon, and McDonalds.

Among the governments that offered helpful features on their websites were Indiana (a text reader that helps visually impaired and foreign language visitors to the site by reading the web page aloud, in English or another language), Michigan (RSS [Really Simple Syndication] feeds, which deliver changing web content directly to users, and foreign language translation), Minnesota (RxConnect, which offers prescription price comparisons and a methamphetamine offender registry), Missouri (Attorney General's office, methamphetamine complaint form), Montana (services accompanied by demos that walk the user through the various steps), Montana (a methamphetamine cleanup program under the Department of Environmental Quality), North Carolina (a Silver

Alert system, under the Department of Crime Control and Public Safety, allowing users to notify the public of missing persons with dementia or other cognitive issues), North Dakota (a portal that sends e-postcards), Wyoming (facilities allowing users to chat with health care providers, view course descriptions and orders, pay tickets, and book tours of the state capitol), and Wisconsin (a business wizard to help users find information on starting a business, an interactive statewide construction map, and a rare mammal observation form).

Privacy and Security

A growing number of sites offer privacy and security statements, yet they remain more prevalent on commercial than on public sites. Ninety-seven percent of corporate sites had a privacy policy, compared with 73 percent of state government sites and 84 percent of federal government sites. Seventy-nine percent of corporate sites have a security policy, whereas only 57 percent of state government sites do and 77 percent of federal sites do.

To assess particular aspects of privacy and security, I evaluated the content of these publicly posted statements. For privacy policies, I looked at several features: whether the privacy statement prohibits commercial marketing of visitor information; whether the site uses permanent cookies or saves profiles of individual visitors; whether the site discloses personal information without the prior consent of the visitor or discloses visitor information to law enforcement agents.

In this analysis, I found that the public sector did a better job than commercial companies in protecting consumer privacy. For example, only 10 percent of corporate sites prohibit the use of cookies, whereas 39 percent of state government sites and 56 percent of federal government websites offer this protection. Ninety-one percent of corporate sites say they share information with law enforcement, compared with 48 percent of state agencies and 72 percent of federal agencies.

Readability

Literacy is the ability to read and understand written information. According to national statistics, about half of the American population reads at the eighth grade level or lower. A number of writers have evaluated text from health warning labels to government documents to see whether they are written at a level that can be understood by most citizens. The fear, of course, is that too many documents and information sources are written at too high a level for citizens to comprehend.

To see how various websites fared, I used a test of the grade-level readability of each website that we studied. We employed the Flesch-Kincaid standard to judge each site's readability level. The Flesch-Kincaid is a standard reading evaluation test and is the one used by the U.S. Department of Defense. Document scores are computed by dividing the average sentence length (the number of words divided by number of sentences) by the average number of syllables in each word (the number of syllables divided by the number of words).

The average grade readability level of corporate sites was at the 12.5 grade (that is, the average reading level of a person halfway through the first year in college). This is higher than the 11.9 grade for state sites and 10.5 grade level for the federal government. Based on our analysis, each of these numbers is well above the reading comprehension of the typical American.

Disability Access

Corporate sites featured lower levels of disability access than the public sphere. We tested disability access by examining the actual accessibility of websites through the Wave Version 4.0 software (http://wave.webaim.org) developed by the Center for Persons with Disabilities at Utah State University. This organization offers software that tests websites against the standards recommended by the World Wide Web Consortium. I used this software to judge whether sites are in compliance with the priority level 1 standards recommended by the consortium. Sites are judged to be either in compliance or not in compliance based on the results of this test. In this study, 16 percent of corporate sites satisfied the World Wide Web Consortium standard of accessibility. This is lower than the 19 percent of state sites and 25 percent of federal sites meeting that standard.

Foreign Language Access

Corporate sites were best at providing foreign language access. Seventy-nine percent of corporate sites offered a foreign language option, compared with 40 percent of state government sites and 43 percent of federal government websites. By foreign language feature, I mean any accommodation to the non-English-speaking visitor, such as a text translation into a nonnative language.

Ads, User Fees, and Premium Fees

Not surprisingly, corporate sites are much more likely to feature commercial advertising. Fifty-six percent of those examined had ads, compared with 3 percent of state government sites and 2 percent of federal sites. When defining an advertisement, I eliminated computer software available for free download (such as Adobe Acrobat Reader, Netscape Navigator, and Microsoft Internet Explorer), which are necessary for viewing or accessing particular products or publications. Links to commercial products or services available for a fee were considered advertisements, as were banner, pop-up, and fly-by ads.

There was little difference between the private and public sectors in employment of user or premium fees. Six percent of corporate sites, 7 percent of state government sites, and 3 percent of federal government sites had user fees. Few sites employed premium fees to access content. By a premium fee, I mean financial charges that are required to access particular areas on the website, such as business services, access to databases, or viewing up-to-the-minute content. A charge is classified as a premium fee if a payment is required in order to enter a general area of the website or access a set of premium services.

Public Outreach

One of the most promising aspects of digital technology is its ability to bring people closer to businesses and governments. In our examination of websites, we determined whether a visitor to the website could e-mail a person, other than the webmaster, in the particular department. In general, most sites had this feature (97 percent of corporate sites, 88 percent of state government sites, and 82 percent of federal websites).

However, corporate sites were much more likely than government sites to have areas where visitors could offer feedback on the organization. Ninety percent of corporate sites had means for making comments on the site, compared to 48 percent of state sites and 62 percent of federal websites. These areas allowed visitors to post comments or use message boards, surveys, and chat rooms.

Corporate sites did a better job of using interactive features. Ninety-eight percent of them allowed citizens to register to receive updates regarding specific issues, compared with 43 percent within state government and 74 percent at the federal level. By entering their e-mail address, street address, or telephone number, web visitors can receive information about a particular subject as new information becomes available. Such information can range from a monthly e-newsletter highlighting an attorney general's recent opinions to alerts notifying citizens whenever a particular portion of the website is updated. There was little difference across sites in their ability to personalize information to the interests of the visitor. Twenty-nine percent of corporate sites allowed this, compared with 25 percent of state sites and 31 percent of federal sites. Ten percent of company sites provided PDA access, a higher proportion than the 3 percent for state government and 2 percent for the federal government.

CASE STUDIES OF LEADING CORPORATE INNOVATORS

It is clear from accumulating research that innovation is important for economic development, efficiency, and effectiveness. The private sector reaped extraordinary benefits over the past two decades in using technology to improve productivity. Indeed, the virtue of the technology revolution is that it allows organizations to gain economies of scale that improve efficiency and effectiveness.

To see what lessons could be drawn from the private sector, our team conducted interviews with leaders in corporations with a demonstrated track record of innovation.[3] These are individuals who worked at companies that scored well in our overall ratings and who were directly involved in overseeing the surveyed online activities. The case studies of successful technology innovation presented below illustrate how these companies innovated, what the keys to success were, and how major obstacles were overcome.

Wells Fargo

Wells Fargo is a leading financial services company with branch offices across the country. Secil Watson, the senior vice president of Internet Services Group, considers the corporation quite "revolutionary" in the extent to which it has reduced the cost of technology in recent years. During the first dot.com boom, she recalls—before the dot.com bust—$5 million in "seed funding" was required to start an online business. Today, that same website can be created for $500,000. With the technology innovations of the past fifteen years, she says, "if you want to start something, it's actually very cost effective."

The Wells Fargo website has been successful not only in its online features but also in its customer satisfaction. Watson credits this to the company's user-centered design process, whereby the company tries to "bring the user to the table at every step of the decision making process." Among other things, this means that a company needs to understand why customers are going to the website and know what kinds of tasks they perform online once they are there. A company must be aware of its "deepest promise" to its customers. "Really delivering on that promise," she says, "is critical and elemental in designing any experience for customers."

One way Wells Fargo tries to understand its customers is through its "voice of the customer" process, which focuses on customer ideas about existing experiences and features. Watson's team looks at call center data to see what customers are calling, writing, and complaining about. Watson thinks that companies should turn customer complaints into opportunities, whether by fixing the issue or creating a new functionality. The Internet Services Group looks at existing survey tools and other social media channels such as blogs and other websites. It "bring[s] all of these things back together, centralize[s] them, and then create[s] a dictionary of all the information, so that [it] can start quantifying and watching trends." By making the data searchable, product managers and others can look for and make sense of the information.

Wells Fargo conducts qualitative customer research through its "corporate ethnographies." Representatives actually visit customer homes and offices to see firsthand "how the customers manage broadly the tasks that concern banking." They ask the customers to keep diaries, writing about their experiences with Wells Fargo services and their reactions to those experiences. These include cross-channel and cross-product interactions, including experiences at a Wells Fargo bank, on the website, or with a customer service representative. As the company likes to have "a significant amount of verbatim information from the customer" when it does surveys, it keeps the questions broad and open ended.

Once group members have the qualitative results, they sit down with a cross-functional team to go through all of the information, sort the information by theme, and brainstorm. This allows the group to build customer profiles or personas, to answer the questions, "Who are our customers?" and "What kinds of things are important to them?" Group members look at behavioral data on customers—what accounts they hold and demographic information—and then quantify these data (for example, this person represents 10–15 percent of our customers).

At the company's quarterly and annual planning cycles, project managers "leverage the insights from the previous six months and use the different tools that they have from customer satisfaction reports, voice-of-the-customer reports, to ethnographies and behavioral metrics." Watson says the project managers look at all of this information and focus on "concept generation and concept development." As a business, of course, they also need to put a business case on the innovation, considering returns and market size. The way Wells Fargo thinks about technology innovation is to consider "What would make the most business sense? Is it the right channel for customers to use? For a bank to offer?"

Watson thinks these efforts to manage the senior leadership's expectations of technological innovation are helpful at Wells Fargo because most of the bank's senior managers have some background in technology. Senior managers unfamiliar with technology might think that the initial funding request for a project is an exclusive and final request. However, as Watson points out, "it's not like that with websites. There's always a tail to the innovation cost—and cleanup costs too, to remove features that are out of date or not frequently used." As such, the technology budgets of both private companies and public agencies need to focus on that "long tail." "Once you build that capability, customers may want it forever and ever and ever." Since the company "can't always build a bigger boat," however, "sometimes, you have to re-architect" what you already have.

Wells Fargo is "very decentralized" in its organization structure. "We put the power into the professionals," she says, so product managers are in charge of their section of their web-site. When asked whether this increased competition for funding and resources, Watson responded, "it's a delicate balance." The company addresses the issue of resource competition by making it clear that product management groups "have to work with the same number of resources" when it comes to innovation. It has done so by designating the insight and experience design groups who work with the teams as "horizontal groups" or "base resources," as Watson calls them. To encourage innovation, Wells Fargo does not ask its teams to "incrementally fund" the experi-ence design team. "If everything is an incremental expense," Wat-son explains, "then people might want to skip it, and discount it." Essentially, if teams themselves have to pay for every step in the innovation process, they might be inclined to bypass some of the steps entirely and go straight to the finish line.

One problem websites often have is that they sometimes try to give too much information. "But people don't go to websites for information," she says, "they go there to do certain tasks." She does not think that a website is about "what's new [or] what's current." She believes that "it's more about 'what's in it for me.'" Translated into website design, this means putting custom-ers "much more in the driver's seat" and allowing them to "find things intuitively."

On the subject of innovating effectively in the public sec-tor, Watson thinks "it's definitely do-able." In some cases, she observes, the public sector "seems to be on the ball," such as in the increasing use of alternative communications pathways. For example, she finds efforts to use blogs or Twitter to get people engaged in public issues very effective. She points to her local California Department of Motor Vehicles as an example of an organization that has done well in taking advantage of technol-ogy. "They actually have a very good website," she says. "You can find all the forms that you need, and you can schedule an appointment." They have "a really good cross-channel process."

Watson emphasizes the importance of "be[ing] clear about who your customers are" as well as the goals the website is trying to achieve. She notes that this can be more difficult for the public sector, which often works with multiple constituencies and multiple goals. She says that it would be difficult for Wells Fargo if the company served lots of constituent groups. Sometimes "you need to make hard trade-offs, in the way you architect the website." The most rewarding efforts are those that "do the most for the most customers, or for the high-value customers."

AT&T

The telecommunications firm AT&T has moved beyond telephone landlines and long-distance service into mobile and wireless communications. According to Phil Bienert, the vice president of ATT.com, the beauty of the word *innovation* is that it opens up "virtually unlimited possibilities." This notion, combined with that of web space, has led to the emergence of concepts that simply did not exist years ago. He credits innovation with "driving this rapid, constant evolution of what's taking place online"—for example, the current social media boom.

Bienert calls technology an "enabler" that allows customers to "find out where their kids are through their mobile devices, or be able to share all their vacation pictures with their friends." As such, he sees innovation less from a technology perspective and more from what customers can do with it. Ideas for web innovation come from a mix of sources—from the leadership and from employees, but "the vast majority of what drives what happens on ATT.com comes from customers." Sometimes, this is in the form of "explicit feedback" like usability surveys or comments left on the website. At other times, suggestions come from direct customer requests or observing what customers are doing on the website.

To gather user information, Bienert relies on web "dashboards," which he looks at several times a day, "watching what takes place on the website in real time and seeing what's happening with every single mobility site." This information is useful in optimizing customer satisfaction with use of the site. Focusing on even "one or two basis points of improvement every week" means constantly "moving the needle."

While ATT.com can compare itself to the "competitive set" in its own industry, its customers are "going to ATT.com, then Yahoo, then Apple," so these are the sites where the company is benchmarking itself. Bienert explains that AT&T is "always looking at what's taking place on the other websites" because that's where their customers' "expectations [of an online experience] are being set."

The AT&T leadership understands the "big picture" and the importance of innovation. "At every step in the chain-of-command, there is a real understanding and appreciation of it." Bienert sees a "huge amount of enthusiasm from senior management," who use "the site themselves, so they can internalize it."

While "getting the mandate has been less difficult" at AT&T than elsewhere, Bienert notes that the company "still ha[s] to get into the nuts and bolts" to be able to deliver and be accountable. "Whatever we do for the site," he said, "we have to show how it's moving the needle for the company." He later added, "Stakeholders are looking at what you're spending, and what you're doing."

Bienert states that public agencies can take a page from the private sector experience in technology innovation. "The principles that make for a great experience apply to any sector." "First and foremost," he argues, an organization must "start with the customers." He would encourage public agencies like the Internal Revenue Service and the postal service to ask themselves, "Who are [our] customers? And what are they trying to accomplish?" and then let answers to those questions "drive [their] definition of innovation."

Government agencies may have a difficult time understanding who their customers really are. The public? Congress? The media? "Not that AT&T doesn't have multiple stakeholders," Bienert notes, but in the public sector, the complexities of having a number of stakeholders "can be distracting in staying focused on what they want to accomplish."

It can be done, though, as is clear from government sites that have done well. Bienert points to NASA as one such example. "The sites that NASA has built, it's clear they started with their customers—the public—in mind," says Bienert. "They have done a phenomenal job of understanding what their audience is looking for." He adds that the IRS has "done a lot of great things with online documents."

One advantage to letting customer needs drive a company's priorities for innovation is that it can help it avoid "making things that are splashy and fancy" in the immediate moment but ultimately unproductive. Bienert strongly advises against taking a superficial short-term approach to innovation. "We can't repeat sins of the last dot.com bust."

Bienert cautions against running into the fallacy–"which agencies and a lot of big companies have done"—that if the investment and the big upgrades are put in, success will happen. "You don't turn . . . around overnight." Creating a good website and innovating effectively "requires investment over time." Bienert recommends having a "a continuous and sustained long-term view on investment" because "the shelf life for something online is very, very short. Something good today, six months from now is stale and old." Successful innovation is a constant process for which "there's no finish line."

Looking ahead, Bienert says that the "mobility factor combined with the trend about social networking" means that technologies are popping up and "evolving very quickly." This has lead to a lot of "noise" in the marketplace, which he thinks is important to distinguish from real trends. The "noise-to-genuine-trend ratio is starting to get out of kilter." He notes that "even doing the most mundane task efficiently is more innovation that anything you can do that is exciting and has video or social networking."

At the end of the day, Bienert explains, "AT&T is a very large company, with very specific objectives on sales and customer service, and they can't afford to derail that by chasing the latest, hottest trend." Still, the company also does not want to be viewed as behind the times and failing to meet customer expectations. "Don't throw your resources at stuff that people will say is 'so 2009,'" advises Bienert, but do pursue "thoughtful innovation that's oriented around customer needs." Companies and agencies should seek to "balance being innovative with not wasting resources on trends that will have a short longevity in the marketplace."

AT&T does this by "feeling [its] way ahead to see where the core of the market will go." The company might test out a feature on a section of its site—the Online Marketplace feature, for example—follow it, learn from it, see how customers use and respond to it, and then integrate the insights gained into the mainstream shopping experience. As Bienert puts it, "There are places where without making massive investments, you can test market reaction."

FedEx

FedEx is a major shipping company with operations around the globe. Russ Fleming, the vice president for digital access marketing, explains that from the outset, the company's founder, Fred Smith, recognized the value technology could bring to business, with its potential to provide expedited, real-time delivered service to customers. Smith viewed technology as the "backbone of business," an insight that has guided the company's innovation. "I think there were some technology innovators, some visionary technologists," Fleming says, "who saw that you could deliver customer service through the Internet, just like you did over the phone." These technologists believed that a website "could be more than brochure-ware, but transactional." What happened next, Fleming notes, was a "marriage of that insight on customer service, with what that new channel could do." FedEx took the technology that was working for operators and applied it to customers. In the six or seven years since the launch of the FedEx site, the company has steadily moved transactions "from 1-800 to web."

Smith directed the company's development with the understanding that "running FedEx would require a big investment in technology." Company spokesperson Matt Ceniceros estimates that FedEx currently devotes a little bit more than $1 billion of its $39 billion revenue to technology innovation. This puts IT spending at 2.5 percent of the company's budget. He said this figure has remained steady throughout the years and is higher than the 1.88 percent average for governments across America.

FedEx redesigns its website every three years. Fleming says that design is always driven by the needs of the business. Since it is not possible to redesign everything at once, the company generally focuses on one country or regional site (for example, North America) first before rolling out the change to other country sites (of which there are more than 200). He says that FedEx has a high-quality internal development team, and so most of its website development is done in-house. However, staff designers do consult third-party experts for design advice as well.

Fleming attributes the success of the FedEx website to the company's strong understanding of its customer base. "We spend a lot of time trying to define our customers," he says, by conducting usability groups, customer interviews, and a variety of other surveys. The aggregation of this feedback provides the company with fuller insight into the customer experience. For example, FedEx learned that customers take *shipping* to mean a number of things, beyond just one simple act of mailing a parcel. In the company's view, shipping involves everything from requesting a pickup, to printing a label, tracking an item, and verifying its delivery.

Understanding its customer base is something else that Fleming believes has served FedEx well in its technology innovation.

By distinguishing between consumers and business professionals and "becoming more sophisticated about user types," Fleming explained, "we can more effectively serve our customers." As evidence of the company's success, Ceniceros points out that in 1996 FedEx received the Malcolm Baldridge Award for customer service based on its ability to deliver for its users.

Fleming sees challenges in the "proliferation of websites and technologies" and in the changing expectations of customers as FedEx continues to move forward. "We're leaders in using a website to enable your business," said Ceniceros, so FedEx is "looking at technologies that are also dot.com, looking at the mobile environment, looking at shipping applications for the iPhone," and using technologies from Adobe that would allow tracking applications to stay live on a desktop. In addition, the company is exploring other avenues to engage customers, including social networks, blogs, and Twitter.

Based on what has worked well for FedEx in technology innovation, Ceniceros recommends the following to public agencies who want to succeed: "Define [your] business needs, get more from [your] investments, look beyond dot.com and CPU-based computing, define the customer base, and compete." Fleming believes that these are areas in which the public sector has not done so well and that have challenged its ability to innovate in technology.

Understanding the makeup of their constituencies is important for government agencies because "if they're really clear about who their customer is, if they have a customer at the center of their universe," then the agencies can focus their efforts on how that particular customer can be served, differentiating the needs of individual customers, says Fleming.

Lack of competition, according to Fleming, is significant for the public sector, because "a competing agency doesn't necessarily preclude you from existing, so you can still get funded by legislation." He considers this an "impediment to finding ways to innovate" and suggests that the public sector could become more innovative if it found ways to "artificially put those measures and indicators in place."

As Fleming sees it, a third challenge to public sector technology innovation is that public agencies' revenue "isn't often determined by how they serve their customer base." Lacking a market response, he believes, agencies should be measured on performance and efficacy "to shore up the fact that they aren't directly driving the revenue that they get." He suggests that agencies could perhaps "identify companies in the private sector that mirror [the agencies'] reason for being" and thereby "establish accountability through a private sector lens."

To overcome the challenges of organizational fragmentation and resource constraints, Fleming notes, an organization must first understand the governance structure that is in place. Then it should "leverage all the tools at [its] disposal," whether research results or customer feedback, and prioritize its initiatives among competing opportunities. When promoting technological innovation, an organization will consider its "relative value to customers, the relative value to the business, and balance the two."

CONCLUSION

Our findings suggest that on most dimensions of technology innovation, the private sector outpaces the public sector. On dimensions such as interactivity, personalization, and language translation, the corporate websites analyzed performed better than the government agencies. However, this was not the case on privacy policies and disability access. In each of these areas, government agencies rated better than their commercial counterparts. In areas where the public sector cares deeply, its sites can outperform their private sector counterparts. Its values are more protective of the broad public interest than those of private corporations.

The key for successful technology innovation in the public sector is to add interactive features, provide for greater customization, and incorporate visitor feedback in agency operations.

Government departments need to become more collaborative in their decisionmaking processes. One of the reasons why the private sector does well on technology innovation is that businesses pay attention to their customers and draw on the experiences and expertise of visitors to their websites. Taking advantage of citizens' judgments is a great way to leverage outside knowledge. Involving more people in key decisions would help the public sector become more innovative and entrepreneurial.

CHAPTER SEVEN

MOBILE COMMUNICATIONS

THE UNITED STATES HAS COME a long way in its communications network since the Advanced Research Projects Agency Network known as ARPANET was developed in 1969. Designed to link four scientific labs for interoffice communications, that network has given rise to personal computers, the Internet, high-speed broadband, and mobile devices.[1] Using these digital platforms, new applications for economic development, communications, education, health information technology, and smart energy grids are emerging with the goal of connecting people, businesses, and governments.

This burst of innovation has created new marketplace opportunities and challenges. Our national leaders are devoting special attention to broadband policy because it is vital to U.S. infrastructure and long-term economic development. Similar to highways, bridges, and dams, communications represent an infrastructure issue that makes it possible for businesses to stay connected, innovate, and create jobs. Just as we need a strong interstate highway system and viable mass transit, we require accessible and affordable broadband so that businesses and consumers can reap the benefits of wireless technology.

What are consumers' sentiments about their cell phones? How do they use them? What do they love about them? What are their frustrations? What changes would they like to see in their service? My researchers and I explored these questions with cell phone users in four countries: the United States, the United Kingdom, Spain, and Japan. These nations represent diversity in terms of geographic location, political and economic development, regulatory regimes, and broadband policies. By evaluating consumer attitudes in places with different characteristics, much can be learned about how government policies affect consumer behavior and how policy changes might affect the commercial marketplace in the future.[2]

METHODOLOGY

For the analysis of consumer behavior and attitudes, I rely on a Zogby International study of 3,428 mobile phone consumers completed between December 2008 and February 2009. This research project surveyed 1,800 cell phone users in the United States, 617 in Japan, 507 in the United Kingdom, and 504 in Spain. The margin of error for the U.S. survey was +/–2.9 percentage points. In Spain and the United Kingdom, the margin of error was +/–4.5 percent; and in Japan, +/–4.0 percent.

The surveys asked identical questions about cell phone usage, whether the customer has downloaded cell phone features, frustrations with current cell phones, and the willingness to pay more to control cell phone features. The American survey added a number of additional questions, including customers' views about what drives innovation, the features they would like to add to their cell phones, the likelihood of switching from landline to cell phone, and the impact on usage of the current economic crisis. Questions on standard demographic items, such as age, income, and race, were asked, allowing me to examine responses by background characteristics.

CONSUMER BEHAVIOR AND ATTITUDES IN THE UNITED STATES, THE UNITED KINGDOM, SPAIN, AND JAPAN

Spain is a country that has embraced open architecture and innovation in the telecommunications field as a matter of national policy. It deregulated telecommunications in the 1990s and ended the monopoly previously held by Telefonica, encouraging the company to expand abroad. Its national plan for research, development, and innovation has budgeted annual research and development expenditures at 1.4 percent of GDP, up from its previous figure of 1.0 percent. It now devotes 2.5 percent of GDP to spending on innovation.[3]

Japan is more closed in terms of mobile communications. It has not deregulated its telecommunications markets and does not make it easy for foreign companies to operate there. The country has not embraced smartphones to the extent of other nations. Its telecommunications network has relatively high consumer prices and low levels of cell phone innovation.

The United States and the United Kingdom are intermediate systems with a blend of service providers, new technologies, and broadband policies. Each has opened its network in certain respects but has been criticized for not going further toward telecommunications liberalization. Some consumers have moved to smartphones, wireless access, and high-speed broadband. But there remain inequities in access based on age, income, and race.

Use of Smartphones

Of the four countries polled, the United States has the highest percentage of cell phone users relying on a smartphone or PDA. Overall, 18.9 percent have smartphones, compared with 12.3 percent in Spain, 7.5 percent in the United Kingdom, and 0.8 percent in Japan. Japan has the lowest rate of smartphone usage because it has not opened its telecommunications infrastructure to foreign companies that make these kinds of devices, and domestic manufacturers have not produced a Japanese version.

This illustrates how the level of competition matters for tele-communications. Consumers in countries that have not liberalized their phone manufacturing policies to allow entry by international companies are less likely to make use of advanced technologies. Ordinary folks are not in a position to take advantage of smart-phones or other advanced forms of telecommunications.

Mobile Device as Phone or as Personal Computer

By a margin of 10 or 20 to 1, consumers overwhelmingly see their mobile device as a phone rather than as a personal computer. In the United States, for example, 77.6 percent consider calling oth-ers as the primary purpose of their mobile device, compared with only 2.8 percent who view it as a computer.

But there are significant variations by country. Spain's consum-ers have the most expansive view of mobile technologies. In Spain, 47.2 percent considered their cell phone an extension of their per-sonal computer, compared with 31.2 percent in the United States, 28.6 percent in the United Kingdom, and 25.5 percent in Japan. The contrasts in these numbers demonstrate the importance of open architecture in the way people define their mobile device. Consumers in countries with more open telecommunications are more likely to see their cell phone as a personal computer because applications are available that allow their phone to function like a computer.

Choosing One's Own Cell Phone Applications

Customers in the United States expressed a strong desire to choose their own cell phone applications. Overall, 80.5 percent of Ameri-cans want this control, compared with 65.7 percent of surveyed users in the United Kingdom, 59.5 percent of those in Spain, and 37.9 percent in Japan.

When asked to identify the most important cell phone features, Americans named chat or instant messaging (23.2 percent), cheap international calling (15.5 percent), file sharing (7.6 percent), and video conferencing (5.2 percent). In the United Kingdom, chat was the top preference (30.4 percent), followed by cheap international rates (29.4 percent), file sharing (9.7 percent), and video conferencing (1.2 percent).

In Spain, file sharing was the most desired application (33.7 percent), followed by cheap international calling (21.4 percent), chat (12.5 percent), and video conferencing (10.7 percent). For the Japanese, cheap international calls were the most favored (12.5 percent), followed by chat (12.0 percent), file sharing (7.6 percent), and video conferencing (1.5 percent).

Downloading Cell Phone Applications

Among the nations surveyed, Spain had the largest percentage of consumers who have downloaded cell phone applications. Forty-eight percent reported having done so, compared with 27.8 percent in the United Kingdom, 26.2 percent in the United States, and 22.4 percent in Japan.

The top reasons users gave for not downloading applications in the United States included lack of interest (37.2 percent), cost concerns (16 percent), not available on the device (12.7 percent), user did not know how to download (3.6 percent), and service provider restrictions (1.1 percent).

In the United Kingdom, the most frequently named reasons were lack of interest (34.1 percent), cost worries (15.6 percent), not available on the device (10.8 percent), did not know how to download (6.3 percent), and service provider restrictions (1.0 percent).

For Spain, the top reasons were cost worries (23.8 percent), lack of interest (20.6 percent), not available on the device (2.6 percent), did not know how to download (1.4 percent), and service provider restrictions (0.2 percent).

For the Japanese, the most common stumbling blocks to downloading applications were lack of interest (36.8 percent), cost (23.5 percent), did not know how to download (4.2 percent), not available on the device (3.7 percent), and service provider restrictions (0.8 percent).

Cell Phone Service Frustrations

The biggest cell phone service frustration in the United States was the length of service contracts. Most telecommunication carriers require extended time contracts (often two years). This locks consumers into particular vendors and prevents them from taking advantage of new opportunities or cheaper rates. Following that frustration were roaming charges (20.4 percent), the cost of domestic calls (18.1 percent), lack of features (15.2 percent), inability to transfer devices across providers (14.8 percent), and lack of interoperability (12.0 percent).

In the United Kingdom, the most common complaints were the cost of international calls (27.8 percent), the cost of domestic calls (25.0 percent), the length of service contracts (22.7 percent), and roaming charges (22.3 percent). For Spaniards, the greatest frustrations were the length of service contracts (41.1 percent), cost of domestic calls (40.1 percent), roaming charges (25.6 percent), inability to transfer devices (22.8 percent), and the cost of international calls (21.6 percent). In Japan, the greatest sources of frustration were the cost of domestic calls (32.3 percent), lack of features (18.0 percent), lack of interoperability (15.4 percent), length of service contract (14.4 percent), and the slow pace of innovation (14.4 percent).

Paying More to Control Cell Phone Applications

Pollsters asked cell phone consumers in each country about their willingness to pay more money for the ability to choose their cell phone applications. The country with the greatest willingness to pay more was Spain (50.0 percent), followed by the United Kingdom

(35.7 percent), the United States (32.9 percent), and Japan (17.2 percent). These numbers suggest that when given more control and greater choices, customers are willing to pay more if they see value arising from their personal investment. Service providers in Spain give consumers extensive options, and this contributes to peoples' willingness to pay more to control their cell phone applications.

DETAILED ANALYSIS OF AMERICAN
CONSUMER BEHAVIOR AND ATTITUDES

The survey taken in the United States included a more detailed questionnaire than that in the other countries. In addition to the items analyzed above, this project queried people's views on tele-communications innovation, the features they would like to add to their cell phone, and the likelihood of their switching from landline to cell phone. Since these items were not asked of cell phone users in the United Kingdom, Spain, or Japan, I focus here on the American responses and describe U.S. consumers' attitude toward telecommunications innovation.

What Drives Innovation

We asked American consumers which of the following they thought represented the greatest source of innovation: new voice services from telephone providers, new Internet-based services such as Google, or new manufacturing devices such as those from Apple, Nokia, and Motorola. American consumers were most likely to see the most extensive innovation from new manufac-tured devices (32.7 percent), followed by new Internet-based ser-vices (28.6 percent), and new voice services (10.4 percent).

When asked what change would be most important to improv-ing their experience with cell phones, users named getting less expensive service from mobile carriers (55.5 percent) as their top preference. They also wanted to see a general-purpose device that would serve all their needs (22.7 percent), to have more control over customizing mobile devices (7.8 percent), to replace their computer with a mobile device (3.3 percent), and to be able to make high-quality, live mobile video calls (1.2 percent).

Popular Downloaded Cell Phone Applications

Games were the most popular application that American respondents had downloaded to their cell phones. Overall, 61.6 percent named games, followed by local directories with information about businesses, restaurants, and the like (52.9 percent), music (49.8 percent), chat (39.8 percent), and social networking (31.8 percent) applications, and video players (19.0 percent).

Switching from Landline to Cell Phone

Nearly half of Americans (48.9 percent) said they were likely to switch from landline to cell phone. Forty-eight percent indicated that if they were planning a short trip of a few days, they would bring along their cell phone rather than their laptop. About one-third (35.4 percent) would discontinue their landline if they had access to a service featuring Google voice innovations.

Impact of Economic Crisis

The current economic crisis has had some impact on consumers. When asked how it had changed their cell phone usage, 5.8 percent said they were making fewer calls, 4.8 percent indicated they had postponed upgrading their phone, 4.4 percent were cutting back on services, 1.2 percent were sending fewer text messages, and 1.0 percent were downloading less content.

DEMOGRAPHIC BREAKDOWNS IN AMERICAN CONSUMER BEHAVIORS AND ATTITUDES

The American survey sample of 1,800 responses was big enough to allow breakdowns by a variety of demographic categories. Using these data, we looked at how people's responses varied by age, race, and income to see the manner in which these background factors affected the attitudes and behavior of cell phone users. A number of interesting differences based on demographics were evident.

Switching from Landline to Cell Phone

Young people in our survey were the most likely to say they would switch from landline to cell phone. More than two-thirds (69.3 percent) of consumers aged eighteen to twenty-nine indicated they would switch, as did 53.9 percent of those aged thirty to forty-nine. Switching became less likely among older groups, but around one-third of those over the age of fifty stated they were likely to switch to a cell phone.

Whites were the most likely racial group to say they would switch. Nearly half (44.1 percent) of Hispanics, 41.8 percent of Asian Americans, and 33.3 percent of African Americans were also willing to make the change.

Nearly half of those consumers earning more than $25,000 said they were likely to switch to a cell phone. The group least likely to make the change were people making less than $25,000.

Choosing One's Own Cell Phone Applications

The desire to choose cell phone applications was strong across all age, race, and income categories. Asian Americans were the most likely to want to choose (85.6 percent), followed by whites (81.5 percent), African Americans (76.1 percent), and Hispanics (73.9 percent).

Downloading Cell Phone Applications

Young people were the consumers most likely to have downloaded cell phone applications. More than 40 percent of those aged eighteen to twenty-nine have downloaded applications, compared with the 32.4 percent of those aged thirty to forty-nine, 15.7 percent of consumers aged fifty to sixty-four, and 6.7 percent of people over the age of sixty-five years who have done so.

African Americans were the group most likely to have downloaded applications (33.4 percent), compared with Asian Americans (26 percent), whites (25.7 percent), and Hispanics (23.0 percent). And people earning more than $100,000 (34.3 percent) were three times as likely as those making less than $25,000 (11.1 percent) to have downloaded cell phone applications.

Improving Cell Phones

The survey taken in the United States also asked questions about how users' cell phone experience could be improved. I broke down people's responses to five categories: adding a multifunction device, getting less expensive service, gaining more control over applications, making video calls, and replacing computers with cell phones.

The people most interested in having a single, multipurpose device were young people, African Americans, and those earning less than $25,000. Hispanics were the most interested in having less expensive service. Asian Americans were the ones most likely to indicate they would like to have a cell phone that could replace their computer.

Popular Downloaded Cell Phone Applications

I found interesting demographic variations in the applications people have added to their cell phones. African Americans were most likely to have added music and chat. Hispanics placed a high value on social networking, and Asian Americans emphasized local directories. Young people (74.1 percent) were three times as likely as senior citizens (28.6 percent) to want to add games to their cell phones. Asian Americans were the consumers most interested in new video applications.

Most Important Cell Phone Features

Chat features were the applications young people said they would most want added to their cell phones, while Asian Americans

116

were interested in making cheap international calls and sharing files. Video conferencing was the feature of greatest interest to African Americans and to people thirty to forty-nine years old.

Paying More to Control Applications

Younger people, African Americans, and poor people were those most likely to be willing to pay more for choice of cell phone applications. Nearly one-third of those aged eighteen to twenty-nine and 36.6 percent of those aged thirty to forty-nine said they would pay more. About half (50.2 percent) of African Americans and a similar number of Asian Americans (53.0 percent) stated they would be willing to increase their payments. On the income dimension, 44.5 percent of individuals making less than $25,000 indicated they would pay more.

CONCLUSION

Based on our analysis of the data, I drew six important conclusions. First, American consumers express a strong desire to choose and control their own cell phone applications. Nearly half of Americans (48.9 percent) are likely to switch from phone landlines to cell phones. Americans believe that innovation is driven by new devices such as those made by Apple and Nokia and the new Internet features pioneered by Google. The most common cell phone frustrations in the United States are the length of service contracts, roaming charges, cost of domestic calls, and lack of desired features. The most popular new cell phone features are games (named by 61.6 percent), local directories (52.9 percent), music (49.8 percent), and chat and instant messaging (39.8 percent).

Second, there are interesting age differences in cell phone attitudes. Two-thirds of American young people (69.3 percent) are willing to switch from landline to cell phone. Young consumers are the most likely to have added games, music, and social networking to their cell phones. They are the most likely to have downloaded cell phone applications and believe that having one general-purpose

117

mobile device would be an important improvement. Young people would like to add chat features to their cell phones.

Third, we found important racial and ethnic differences in the U.S. survey. African Americans are very interested in adding music to their cell phones, while Hispanics place a high value on social networking and Asian Americans emphasize local directories. Asian Americans are the group that says it is most interested in making cheap international calls.

Fourth, the American groups most willing to pay more to control their cell phone applications are young people, African Americans, Asian Americans, and those earning less than $25,000.

Fifth, reflecting Spain's emphasis on telecommunications reform and open architecture, Spanish consumers are the most likely of the four national groups to see cell phones as an extension of personal computers and the most likely to have downloaded cell phone applications. They are also the group that expresses the greatest willingness to pay more for new cell phone features.

Finally, in each of the four countries, the most popular cell phone features are the ability to chat and make cheap international calls. The most frequent reasons for not downloading new cell phone applications are lack of interest, cost, and user inability to download new applications. The cell phone factor considered most important is less expensive services from mobile carriers.

Speaking more generally, our results demonstrate the virtue of innovation for communications policy. When people are given choices, they act on those options, use services more often, and are willing to pay more because they see the value of their personal investment. Some have suggested that the interaction between telephone service and new Internet providers is a zero-sum game. With cell phone penetration at 90 percent, analysts claim the cell phone market is saturated and not likely to experience future growth.

However, this conclusion ignores the possibility of additional growth if consumers have more control over their applications and expand their usage levels. Data from this project suggest that there are features consumers would like to add to their mobile devices. Consumers would like to have greater control over their applications, and if offered more features they would expand their usage levels and be willing to pay more for services. In this situation, opening the networks would be win-win for businesses because the telecommunications pie would grow larger.

This finding is especially relevant in regard to the leading generation of young people. They are the age group with the most positive attitude toward cell phones, the most likely to have downloaded applications, and the most interested in controlling their applications. They are most comfortable with digital technology and mobile communications. They understand the convenience of new applications and say they would be willing to pay for valued features.

As the United States undertakes new investments in education, health information technology, and energy efficiency, it is important to have broad access to telecommunications and high-speed networks. New platforms will spur usage and innovation and bring additional people, businesses, and services into the digital revolution.

As to future policy decisions, several topics warrant additional attention. First is the need for more extensive data on consumer and business practices. Right now, the Federal Communications Commission lacks detailed information on consumer and business usage. The FCC undertakes a biannual survey of broadband Internet service in various areas across the United States, broken down by zip code. However, this survey does not distinguish the number of providers or report many details on broadband speed or usage. This lack of information makes it hard to judge the current state of affairs and thereby hinders policymaking.

There are some other sources of information. For example, the U.S. Census Bureau undertakes periodic updates on broadband usage through its Current Population Survey. But these data collections have been completed only four times in the past decade (in 2000, 2001, 2003, and 2007) and therefore do not offer sufficient data to inform policymaking.

A couple of states, such as Kentucky and North Carolina, collect more detailed data on broadband usage within their borders. They compile more specific information in terms of how people are using broadband and in what geographic areas. These surveys could represent a model of the kind of information that would be helpful to national policymakers.

Second, the United States needs to determine which policies make the most sense. In the communications area, there are many moving parts: consumer behavior, social background, business practices, regulatory regimes, and legislative policies. It is hard to disentangle the parts to know what policy changes will lead to the greatest innovation and economic development. With more detailed data, though, it would be possible to sort out the linkages and determine which policies would have the greatest impact.

It is clear from the analysis of consumer behavior data that policies have differential effects on various demographic groups. Young people are the most willing to experiment with new technologies and to want more applications. Since they are the rising generation, close attention to how changes in policy affect their use of technology would suggest fruitful directions for future innovation.

PART FOUR

NORMATIVE CONCERNS

CHAPTER EIGHT

PERSONALIZED MEDICINE AND HEALTH INFORMATION TECHNOLOGY

FEDERAL OFFICIALS ARE PURSUING THE goal of a personal human genome map at a cost of less than $1,000 in five years.[1] Thus it is possible to envision a future in which treatments are tailored to individuals' genetic structures, prescriptions are analyzed in advance for likely effectiveness, and researchers study clinical data in real time to learn what works. Implementation of these regimens creates a situation in which treatments are better targeted, health systems save money by identifying therapies not likely to be effective for particular people, and researchers have a better understanding of comparative effectiveness.[2]

Yet despite these benefits, consumer and systemwide gains remain limited by an outmoded policy regime. Federal regulations were developed years before recent advances in gene sequencing, electronic health records, and information technology. Because scientific innovation is running far ahead of public policy, physicians, researchers, and patients are not receiving the full advantage of latest developments. Current policies should leverage new advances in genomics and personalized medicine to individualize diagnosis and treatment. Similarly, policies creating incentives for the adoption of health information technology should ensure that the invested infrastructure is one that supports new care paradigms rather than automating yesterday's health care practices.

To determine what needs to be done, a number of key leaders from government, academia, the nonprofit sector, and business were interviewed about ways to promote a better use of health information technology to enable personalized medicine. The interviews focused on policy and operational issues surrounding interoperability, standards, data-sharing protocols, privacy, predictive modeling, and rapid-learning feedback models.

Enabling personalized medicine and the policy and operational changes that would facilitate connectivity, integration, reimbursement reform, and reanalysis of information face a number of challenges. Our health system requires a seamless and rapid flow of digital information, including genomic, clinical outcome, and claims data. Research derived from clinical care must feed back into assessment to advance care quality for consumers. Discrete data on diagnosis, treatment, medical claims, and health outcomes exist in parts of the system, but it is hard to determine what works and how treatments differ across subgroups. Changes in reimbursement practices would better align incentives with effective health care practices.

We need privacy rules that strike the right balance between privacy and innovation. These rules should distinguish health research from clinical practice and create mechanisms that can connect data from multiple sources into databases for secondary research usage and population cohort analysis. More-balanced rules would improve innovation. It is nearly impossible to evaluate treatment effectiveness without being able to aggregate data and compare results. Faster knowledge management would enable rapid-learning models and evidence-based decisionmaking on the part of physicians and public health officials.

As more information on treatment, lab tests, genomics, and financial costs gets integrated into health care, it is hard to incorporate data from medical history, vital signs, genetic background, and lab testing into diagnosis and treatment. Predictive modeling represents a way for physicians to move toward systematic and evidence-based decisionmaking. The first step toward enabling personalized medicine is ensuring that clinicians have access to what is known about patient gene variants, but computer models can go beyond this approach to predict what treatments are likely to be most effective given observed symptoms. Public policy should incorporate rapid-learning and predictive modeling to gain the full benefits of personalized medicine.

Personalized medicine can be enabled in several ways: meaningful-use requirements promulgated by the executive branch, change driven by consumer demand for personalized medicine, pilot and demonstration projects supported by the Centers for Medicare and Medicaid Services (CMS) Innovation Center, and collaborations between academia and industry encouraged by the government through investment. The declining costs of DNA sequencing will drive consumer demand and generate a growing need for physicians to personalize medicine. In addition, CMS should deploy some of its $10 billion in pilot project resources through its new innovation center to encourage personalized medicine. Along with the National Institutes of Health, the agency could fund new projects designed to demonstrate innovation in health care.[3]

THE CHALLENGES OF ENABLING PERSONALIZED MEDICINE

There are a number of policy and operational challenges that interfere with the public's ability to gain the benefits of personalized medicine through health information technology.[4] These include issues such as interoperability, inconsistent coding and language standards, problems in data sharing, weak feedback loops, privacy concerns, and ineffective reimbursement policies.

Interoperability represents a major challenge because of the difficulty of connecting computing systems and integrating data from different sources. When researchers are not able to exchange information, the cost of health care rises, and learning in real time becomes difficult. A considerable amount of medical information is collected, but too little of it is integrated or put into databases that are usable for research or public health purposes.

Legitimate concerns over privacy and confidentiality complicate secondary use of health care information. Even when data are aggregated and depersonalized, researchers find it hard to gain access to information that helps gain insights into public health trends.

There also are problems in terms of reimbursement policies. Many programs are not well aligned with laudatory goals such as preventive medicine or positive health outcomes. This mismatch makes it difficult to judge quality or build incentives for healthy outcomes. We need to reward providers for good behavior and reduce incentives for wasteful or unnecessary treatment.

THREE REVOLUTIONS AND HOW THEY AFFECT HEALTH CARE

There are three revolutions taking place in health care: major shifts in the delivery of medical care from hierarchy and physician dominance to models featuring a variety of medical providers; a dramatic increase in the amount of digital information available to consumers, and improvements in understanding of human genetics and gene behavior. In this section, I review these dramatic changes and note how they make it possible to personalize medical care in ways never before possible.

The Medical Delivery Revolution

Health care is shifting from a hierarchical delivery system to one that features greater transparency, new providers, and more patient involvement. For much of the twentieth century, medicine was dominated by physicians with considerable professional autonomy, hospitals, the pharmaceutical industry, insurance companies, and government agencies that focused on the elderly, veterans, and the poor.

Now we are seeing the emergence of customer-driven medicine that has expanded the range of nontraditional health care providers and placed more information-gathering responsibility on patients and their caregivers, such as people with elderly parents. Businesses such as CVS and Wal-Mart have developed in-store treatment centers.[5] Out-patient facilities have proliferated at a rapid rate. Patients can order drugs through Internet sites. Rather than rely only on doctors, consumers can get health information from the Internet, social networking sites, fellow patients, and chat rooms.[6]

Remote monitoring devices and mobile health applications allow people to monitor their own weight, blood pressure, pulse, and sugar levels and send results electronically to health care providers. Patients can store their medical records online and have access to them regardless of their geographic location. Some get personalized feedback via e-mail reminders or other types of electronic reminders when they gain weight, have an uptick in their cholesterol levels, do not take their medicine, or see a rise in their blood pressure.

Scans and imaging have improved to a high level of resolution. Imaging tests, especially computed tomography (CT) scans, can measure tissue down to one-third of a millimeter in size. This development allows health care providers to describe symptoms with tremendous precision and monitor patient responses to various therapies. Imaging enhances medical personalization and tailors treatment to individual circumstances.

The Digital Revolution

Concurrent with major changes in medical care delivery has been an explosion of digital resources for patients as well as physicians. There are websites that evaluate hospitals on health outcomes and patient perceptions. In addition, there are social networking sites that share information on chronic disease conditions.

Through these and other digital resources, doctors and patients have much more information at their disposal.[7] They know more about their own histories, can link to additional sources of information, and can interact electronically with health care providers. This strengthens their access to information and encourages them to ask more informed questions about their medical conditions.

As part of the 2009 American Recovery and Reinvestment Act, Congress authorized $44 billion in public funding of physician and hospital adoption of electronic health records. Policymakers hope to extend the use of electronic health records by providing grants to hospitals and physicians meeting meaningful-use standards. Their goal is to increase the use of health information technology from 10 percent to 90 percent of health care providers, indicating that they have adopted electronic health records in meaningful ways, by 2015. The new investment creates the opportunity to adopt information systems that accelerate personalized medicine as opposed to merely automating systems designed years ago.

The Genomic Revolution

Scientists have made extensive progress over the past two decades in understanding human genetics and the role of proteins and chemicals in gene behavior. In 1989 the National Institutes of Health launched the Human Genome Project in an effort to identify the basic building blocks of the human organism. By 2003 investigators had sequenced the genome and identified 3 billion discrete "chemical units."

Since that time, scientists have worked to establish links between gene structures, human illnesses, treatment effectiveness, and adverse effects.[8] Integrating genetic sequencing data into electronic health records potentially cuts health care costs through more effective targeting of treatments and more accurate diagnoses. This type of connectivity speeds research feedback into clinical care and delivers more timely information to patients, physicians, and medical researchers.

Advances in DNA sequencing have made it possible to develop greater understanding regarding the role of genetic structures in disease susceptibility and drug efficacy. Scientists have identified genes that increase the odds of getting illnesses such as breast cancer, cystic fibrosis, and sickle-cell anemia.

For example, they have found that those carrying certain mutations in the BRCA1 or BRCA2 genes have a higher risk of breast cancer, and those with an overexpression of the HER2 protein are at greater risk of reoccurrence. Combined with detailed family histories and diagnostic tests, doctors can pinpoint who is most susceptible to breast cancer and therefore should be monitored most carefully. It is now understood that patients cannot be treated based on population averages; subgroup and individual differences have to be factored in.

Investigators have made progress in determining who is most likely to benefit from possible treatments. In oncology, for example, pathologists measure estrogen receptor expression to determine eligibility for tamoxifen hormone therapy among those suffering from breast cancer. Effectiveness has been found to be contingent on a cytochrome enzyme P450 2D6 needed to metabolize the drug, although the results have not been consistent across studies.[9]

Genetic tests for HLA-B*1502, a particular variant of human leukocyte antigen, are already available and can predict increased susceptibility to dangerous or even fatal skin reactions such as Stevens Johnson syndrome and toxic epidermal necrolysis. These conditions are caused by the carbamazepine therapy used in the treatment of seizures. This allele occurs almost exclusively in patients with ancestry across broad areas of Asia.[10]

There has been mixed evidence regarding a link between the genotype CYP450 and antidepression treatments using selective serotonin reuptake inhibitors. Some patients suffering from metastatic colorectal cancer whose tumors have a gene mutation called KRAS have not responded well to treatments using panitumumab or cetuximab.[11]

Analysis has demonstrated that many patients are not able to benefit from particular drug therapies. For example, scientists have discovered that some cancer treatments are ineffective for 75 percent of patients. Gefitinib (Iressa) and erlotinib (Tarceva) are drugs for treatment of non–small cell lung cancer, but they are effective only for tumors that express the epidermal growth factor receptor gene. Other prescribed medications are ineffective for 70 percent for Alzheimer disease sufferers, 50 percent for those with arthritis, 43 percent who are diabetic, 40 percent who suffer from asthma, and 38 percent who take SSRI antidepressants.[12] Since people metabolize medicine in so many different ways, depending on their particular combination of genes, enzymes, and proteins, genomic information is vital to determining the optimal therapy.[13]

POLICY CHALLENGES AND RECOMMENDATIONS

The convergence of these revolutions in medical delivery, digital technology, and genomics has the potential to generate important benefits for patients, physicians, and public health officials. But to take advantage of these developments, it is vital to connect genomic and other personalized information to electronic health records. If this information is not readily accessible to diagnosing and treating physicians, their ability to coordinate patient care effectively is limited. As researchers learn more, information on genetic history and susceptibility to drugs and side effects needs to be at the fingertips of doctors in the same way that family history, vital signs, and medical tests are. Timely information would help caregivers incorporate what works and does not work into their clinical decisions.

Data-Sharing Networks

One of the biggest barriers to gaining efficiencies in our current system is interoperability problems in connecting different information systems. With 650,000 doctors and 5,800 hospitals, the United States has a health care system that is quite fragmented.

The clinical records of patients do not travel with them electronically, and most of the computing systems do not enable data flow. Many have different systems for compiling billing, lab tests, medical records, prescriptions, treatment, and appointments, which makes it difficult for providers to exchange information outside of electronic converters. To make matters worse, even when integrated the information captured in some of these systems is meaningless because it relies on different semantics that make cross correlation nearly impossible.

Discrete data sets exist in parts of the present system, but they are not integrated. It is hard to determine what works and how to assess costs and benefits. Technology has been used to improve the accounting and administrative aspects of health care but not its knowledge management. We need information systems that aid in the analysis of the overall contours of health care.

In the medical area, the creation of national drug codes created reimbursement efficiencies. Establishing a ten-digit code for each medication helped to make drug administration safer and more economical. It facilitated the tracking of pharmacological information and produced benefits for both consumers and businesses.

Health information technology and electronic health records can serve the same type of integrative role. It is possible to track claim receipts in real time. Coverage effects can be combined with evidence on medical tests and clinical performance, leading to shorter evaluation cycles and an enhanced ability to control costs in ways that do not weaken quality. Treatment guidelines in electronic health records would help physicians understand their treatment options.

The goal of data-sharing networks is to develop a "virtuous cycle" for health care, wherein improvements build on one another. Such networks would require properly coded electronic health records whose data would inform clinical care and evaluate substantive value. Treatment information should be linked to outcomes, and reimbursements based on the end result. A balance between costs and benefits would help people make informed decisions. Right now, there is much greater concern about the costs and burdens of integration than its possible benefits to patients, physicians, researchers, and public health administrators.

An excellent example of a new kind of data exchange is the Cancer Biomedical Informatics Grid, or caBIG. This network, launched in 2003 by the National Cancer Institute, connects more than fifty institute-designated cancer centers, along with other academic and commercial organizations, making it the largest national biomedical information network in the United States. Capabilities compliant with caBIG interoperability specifications enable the collection, analysis, and exchange of a wide range of biomedical information through a well-integrated, standards-based infrastructure coupled with open-source and commercial software applications. These technologies create an integrated electronic system that enables clinical research, genomics, medical images, biospecimens, and patient outcomes data to flow easily but securely among authorized individuals, organizations, and institutions. These capabilities enable health care providers to leverage resources developed in research settings to identify subgroups of molecularly related patients, collect and view patient histories individually and in the aggregate, and collaborate across organizations to test research hypotheses and evaluate new treatments.

Semantics and Data Coding

Current electronic health systems have features that undermine connectivity. Researchers, clinicians, and industry employ inconsistent standards in how medical terms are defined and applied to health conditions. They do not classify diseases in the same way or describe symptoms with similar language. These semantic inconsistencies make it difficult to populate electronic records with data that are comparable.

In the world of specialty care, this problem becomes even more serious. Health providers from various specialties require different information and often record symptoms in dissimilar ways. Pathologists may have informational protocols that are different from those of oncologists or internal medicine physicians. As long as semantic inconsistencies are present, it will be hard to take full advantage of digital record-keeping systems.

Right now, the United States does not have adequate diagnostic coding protocols. Each of our 5,800 hospitals has its own nomenclature and fields of description. Many providers use different terms to describe the same symptoms. The data are overly aggregated, and it is therefore hard to determine what actually is going on. For example, there now are sixty different types of leukemia, but the International Classification of Diseases codes do not reflect the diversity of that disease. Physicians and administrators say they need greater granularity in the coding conventions. Since diagnostic tests represent up to 70 percent of physicians' core decisions, according to informed experts, the best way to evaluate costs involves greater precision in coding lab tests. However, with the future implementation of ICD-10, a more detailed disease classification system, additional granularity with regard to specific diagnosis will be enhanced.

The same problem occurs in regard to genomics. There are no differentiated billing codes for various molecular or genetic conditions. Many health care systems do not distinguish gene tests for various illnesses. This makes it impossible to aggregate data or link genomic information to disease diagnosis and treatment.

There have been improvements on some of these dimensions. The Systematized Nomenclature of Medicine Clinical Terms has developed a disease categorization nomenclature that is used in fifteen countries. But many of its codes are not detailed enough for research purposes. The incompatibility of codes between clinical and research communities prevents each from building on the work of the other. Common points of reference are required that make use of new research as it develops, including more variegated descriptors, more detailed codes, and identification of different types of cancers. The system needs to be dynamic in nature so that it is regularly updated as researchers develop new knowledge about medical illness.

Professional networks are making progress on coding and language description. The role of the Clinical Data Interchange Standards Consortium has been helpful, and its reliance on professional experts from several fields led to the development of widely used standards. Clinical genomics guidelines for health care providers would be useful in providing an overall framework for integrating genetic information into electronic health records. There remains a need for greater specification of different types of genetic background.

Of course, standardized vocabularies work only when there is shared understanding of what the individual words mean. Medicine still is constrained by the imprecision of human language and the vagueness of patient symptoms. True semantic uniformity may come best with machine-generated data rather than perceptual data.[14]

Balancing Privacy Rules

The rise of personalized medicine raises the stakes on privacy considerations. Genetic information, by definition, is deeply personal since genotypes, enzymes, and proteins are unique to each individual. If released publicly, this information has ramifications for possible employment, economic prospects, and social relationships. If employers knew a prospective employee carried a gene that seriously increased the odds of a chronic disease, would they hire that person? Knowing the benefits of genetic testing for diagnosis and treatment does not mitigate against the possible risk of privacy violations.

There is no question that strong privacy rules are required. People fear discrimination or adverse job consequences from medical information whose confidentiality has been compromised. In part, this is why Congress in 1996 adopted the Health Insurance Portability and Accountability Act (HIPAA). That legislation was designed to address patient privacy concerns and ensure that appropriate safeguards were put into place. The American Recovery and Reinvestment Act has strengthened these privacy rules. The Department of Health and Human Services has a proposed rule out for public comment that would apply HIPAA rules to business associates of covered entities.

In 2008 Congress passed the Genetic Information Nondiscrimination Act prohibiting the use of genomics information in employment hiring or firing. This bill prohibits health insurers from using genetic testing to determine rates and helps to reassure consumers that genomic information cannot be used against them.

However, some experts question whether current privacy rules strike the right balance between privacy and innovation. A 2009 Institute of Medicine report concludes "[that] the HIPAA Privacy Rule does not protect privacy as well as it should, and that, as currently implemented, it impedes important health research." The report suggests people should distinguish "information-based research" from "interventional clinical research." For purposes of clinical research, project analysts argued that it was not feasible to get consent for all secondary uses, especially in situations where the data were "de-identified."[15]

The authors of the report suggest the need to revise privacy rules to distinguish health research from practice and recommend creation of a "mechanism for linking an individual's data from multiple sources, such as databases. so that more useful databases can be made available for research in a manner that protects privacy, confidentiality, and security." In its conclusion, the report calls for a new approach to privacy, saying "effective privacy protections must be implemented in a way that does not hinder health research or inhibit medical advances."[16]

New Approaches to Privacy and Access Control

The question for the health community is how to protect privacy and provide mechanisms for patients' control of access to their medical records. Current limits on data sharing and secondary analysis of de-identified data make it difficult to get the benefits of genomics. Data access is a question not just of the risks of unwanted information releases but also of the benefits that arise from data sharing. For example, it is nearly impossible to evaluate treatment effectiveness without being able to aggregate data and compare results. Researchers need to be able to reuse information so that data can be employed to improve health care quality and cut costs.

Particular privacy issues arise with tissue samples and genomic information. Some experts have suggested that as few as thirteen specific genetic features can identity a particular person with one-in-a-billion certainty. Certain molecular genotypes are rare enough that they can identify a specific individual. Indeed, de-identified qualities nearly evaporate under the notion of personalized medicine.

It is hard to de-identify data when information is specific to one person. Pathologists have a problem with patient consent for future use of a sample because they can never know all the future uses of data for research purposes. In addition, de-identified tissue samples reduce the value of the specimen because the very value of the sample lies in linking it to family history and genetic information. Under current privacy rules, which require that patients approve any current or future use of the tissue sample, it is impossible to gain any of the benefits of genomics.

Consent is complicated when a patient is unsure of future uses of samples. The Havasupai tribe recently resolved its legal dispute with Arizona State University over genetic data gathered from tribal members. University researchers used the material for research into schizophrenia and the tribe's ancestral heritage. Tribe members sued the university over what it claimed were vague consent forms. The two sides settled the dispute, and the university returned blood samples to the original providers.[17]

The Centers for Pathology and Oncology Informatics and the Center for Pathology Quality and Healthcare Research at the University of Pittsburgh Medical Center have developed a novel approach to patient consent on tissue samples. Researchers who need information for study purposes can go to individuals designated by the center's institutional review board as honest brokers. Under particular circumstances, they can request more individualized records to facilitate specific research projects. This procedural mechanism represents a creative attempt to reconcile privacy and research utility.[18]

Greater flexibility by local institutional review boards would help expedite the integration of personalized medicine with health information technology. These boards need to develop new models of real-time testing of cognitive systems. Its members should understand the value not just in assessments of before and after interventions but also in real-time data collection and analysis. It is hard to get human consent in large-scale research projects, but there is value in data aggregation with de-identified information. In this situation, original data can be employed to find unobserved patterns. The secondary use of data for research and public health purposes needs to be recognized as legitimate by local institutional review boards.

In addition, analyses would be helped by research and public health reporting exceptions. One such exception could be limited-use agreements for research purposes that allow scientists to access data for purposes of comparative effectiveness. This could involve removing certain highly sensitive fields from the data set. Administrators should not have to get new permission from patients to report aggregated and de-identified data results to health officials. There is a model in the case of the U.S. Census in which personal information collected from individuals is protected, but aggregated data available down to the census tract and block levels that do not reveal individual identifiers are available to researchers.

There should be better understanding of ways to balance the competing needs of patients, physicians, researchers, and public health administrators. Each requires different kinds of information. For example, patients need access to medical history, vital signs, lab tests, and genomic risks. Physicians need all those things plus genomics, drug interactions, side effects, and clinical treatment guidelines. Researchers need de-identified data on medical history, vital signs, lab tests, genomics, and health outcomes. Public health administrators need data on treatments, diagnosis, financial costs, and health outcomes.

Harmonizing State Laws

Individual states have different rules and requirements on genomics that complicate the administration of medical care, and the HIPAA privacy rules act as a floor, not a ceiling, for stricter state laws. For example, states look for different genetic disorders in newborn screening tests. This lack of continuity makes it difficult to share information across state lines for institutions operating in various jurisdictions.

There also are substantial differences in state privacy and consent rules. A Massachusetts law has been interpreted to outline a distinction between tests done for screening purposes and those for diagnostic purposes. Ordering physicians are not able to perform genetic screening tests without the specific written consent of the patient, and test results are available only to the ordering physician.

This means that other clinicians involved in the patient's care may not have access to this information and may end up making incorrect decisions. For example, a clinician who is not aware that a patient has a genetic predisposition for hypertrophic cardiomyopathy may incorrectly diagnose the patient's symptoms as asthma, with disastrous health consequences. Florida has stricter rules and processes for informed consent. National health care organizations that share data across state borders must be cognizant of the rules in particular jurisdictions and how various regulations affect their mode of operation.

It is common for prognostic gene-testing services to encounter different laws in various states. This affects how much access emergency room physicians have to genomic information. Without access to screening tests or genetic background information that assess the appropriateness of medications such as the anticoagulant drug warfarin, they may administer an ineffective treatment or prescribe the wrong medicine. A doctor might confuse asthma with heart thickening problems, because they have similar symptoms, and then use the wrong medication to treat the condition. This is not only wasteful, it is dangerous to patients.

Greater harmonization of state laws would be helpful to businesses and researchers operating across state lines. It also would speed the development of the IT infrastructure required to support personalized medicine by ensuring that resources are not diverted to the peculiarities of individual state laws. The current cacophony of rules and regulations makes it difficult to know which rules to follow and which regulations apply in cross-border operations.

Ending the Catch-22 of Reimbursements

Reimbursement policies need to be better aligned with national goals. More attention should be devoted to preventive medicine, positive health outcomes, and reimbursement for performance rather than volume of business. Value-based reimbursement would give physicians and hospitals a greater incentive to focus on generating good health as opposed merely to ordering tests.

In 2009 the Centers for Medicare and Medicaid Services investigated whether the federal government should reimburse the costs of genetic testing of patients taking the anticoagulant drug warfarin to determine their likelihood of abnormal metabolism and health care risk. After extensive discussions, CMS decided not to provide blanket coverage of the genetic test because the dosing guidelines and pharmacogenomic association had not been clearly demonstrated.[19]

In recognition of the possibility that genetic background might be relevant for treatment outcomes, CMS did establish a process for test reimbursement if the test were part of a large-scale randomized clinical trial research project. Its thinking was that the patient needed to be enrolled in a formal assessment to qualify for gene test reimbursement. This is an example of how reimbursement policies need to change. This is a classic catch-22: insurers do not reimburse for genetic tests unless there is demonstrated evidence of their value, yet such information cannot be obtained without test performance.

At the same time, current coding protocols do not support the case for genetic information because they do not distinguish various types of genetic tests. This weakness makes it impossible to establish the efficacy of genomic information for medical treatment. Reimbursement policies clearly need to change so that financial incentives are better aligned with policy goals and desired outcomes.

In the case of genetic testing, it is important to distinguish between reimbursements for performing the test and for interpretive services. The physical test of sequencing the base pair and running a DNA analysis is reported on the clinical laboratory fee schedule by a hospital or laboratory and does not include the professional work to interpret and make sense of the data. Some insurers are starting to reimburse for the actual DNA test.

However, physicians and qualified health care practitioners who interpret test results and help patients understand the significance of the information are inadequately reimbursed. In many cases, they are paid less than $10 for each report. With the costs of genetic testing dropping rapidly, the cost of interpretation is becoming a bigger share of the genomic analysis. This situation creates bad incentives in the reimbursement process. It leads health care providers to repeat tests to cover interpretation costs.

Rapid-Learning Feedback Mechanisms in Clinical Care

One of the biggest current problems is the isolation of research from clinical care. It is difficult to integrate advances in medical research into treatment guidelines that are accessible only to physicians. There need to be ways to speed up the feedback loops so that new discoveries get incorporated into treatment.[20]

Administrators collect an extraordinary amount of information, but too little of it is put into a form that is usable for secondary purposes. For example, millions of medical claims are filed, but there is no way to determine which specific tests have been performed. This makes it impossible to identify what lessons can be learned from the data.

Feedback from research to clinical care is weak, and thus it is hard to incorporate the latest research findings into clinical settings. Researchers do not like to share data before publication, and there is great emphasis on the risks of information sharing and little attention to possible benefits of system integration.

Faster knowledge management enables rapid-learning models and evidence-based decisionmaking on the part of physicians and public health officials. Integrated databases help researchers log on and test ideas. High-speed networks, integrated data sets, and research registries facilitate hypothesis testing and allow clinicians to draw on the latest knowledge about medical treatment.[21]

In the long run, it would be valuable to have information both on clinical care and financial management. Integrating cost information into clinical data would allow caregivers to determine the best clinical options at the best cost. Of course, it is notoriously difficult to evaluate medical cost information. Because hospitals provide discounts and rebates, there are proprietary considerations in terms of cost material. Few patients pay list cost for treatments, and Medicare often reimburses only 60 percent of actual costs. In this situation, standard accounting codes that help others determine which payees get which discounts are hard to acquire.

Predictive Modeling in Physician Practices

As more information on treatment, lab tests, genomics, and financial costs get integrated into health care, a major problem arises for physicians. It is hard to incorporate information from medical history, vital signs, genetic background, and lab testing into diagnosis and treatment when there is a flood of information. Knowledge integration may create systems that exceed the cognitive capabilities of care providers to interpret the information. With 7,900 current procedural terminality codes and 80,000 diagnostic codes, physicians and public health professionals find it difficult to figure out what to do with all the information.

Predictive modeling represents a way for physicians to move toward systematic and evidence-based decisionmaking. Computer models monitor the range of information and predict what treatments are likely to be most effective given the range of symptoms observed.

Aided decisions can help modelers develop decision trees for physicians in the style of "what if" questions. If patients have certain symptoms, then what are the logical tests to run and treatments to suggest? Patient- and lab-reported information on symptoms, value statements, comorbidities, vital signs, and renal and heart performance still are central to diagnosis. With the addition of biomarkers and gene testing, predictive models help physicians decide the best course of action. The U.S. Department of Health and Human Services hopes to add learning models to its last stage of financial incentives for health care providers engaged in meaningful use of certified systems. This requirement will enable health information technology to integrate clinical, outcome, and research information.

One example of this approach is Kaiser Permanente's Archimedes model.[22] This approach uses data on blood pressure and sugar levels to chart optimal health. The compilation of information allows computers to locate the individual outliers and determine the probability of illness developing, given patients' life histories and genetic makeup.

Breast cancer is an area that has developed effective predictive models. It is the area in which scientists have made the most progress in identifying biomarkers and determining how to personalize treatment to a patient's genetic composition. It is the leading edge of innovation in personalized medicine.

Questions remain, however, about how best to validate models. Scientists need clinical diagnosis and treatment data linked to health outcomes to determine effective treatments. Approaches that allow medical professionals to learn over time, adapt to new research and new procedures, and feed new information into predictive models represent a promising way to improve health care.

Duke University partners with 100 institutions in an effort to collect the same data from every patient before and after treatment. They follow eighty different data points and employ that information to make decisions on interventions and treatments. They monitor self-reported symptoms and personalize the treatment based on clinical and psychological needs.

Geisinger Health System has used material on patient background and symptoms to predict congestive heart failure disease six to twenty-four months in advance for half of its members who developed the condition.[23] The model was validated based on probabilistic reasoning: If a person has certain symptoms, there is a certain probability that he or she will have heart failure. This helps physicians make decisions on diagnosis and treatment. Public health officials need to incorporate predictive models in physician practices.

CONCLUSION

Better use of health information technology to enable personalized medicine could be promoted in a number of ways. Policy and operational changes could facilitate connectivity, integration, reimbursement reform, and secondary analysis of information. A seamless and rapid flow of digital information, including genomic, clinical outcome, and claims data, could improve both diagnosis and treatment. Research that studies data in real time expedites learning and helps determine what works and how physicians can tailor treatments to individual circumstances.

There are eight changes that would enable personalized medicine:

—creating meaningful-use rules in the Office of National Coordinator of the Department of Health and Human Services that facilitate more effective use of health information technology for personalized medicine

—reducing the isolation of health research from clinical practice and creating mechanisms that connect information from multiple sources into databases for secondary use

—developing privacy rules that strike the right balance between privacy and innovation

—creating differentiated codes for various molecular or genetic tests so that researchers can link genomic information to disease diagnosis and treatment

—building data systems and language semantics that help researchers compare and evaluate information

—enabling feedback loops and creating faster knowledge management to enable rapid-learning models and evidence-based decisionmaking on the part of physicians and public health officials

—deploying predictive models in medical practices to help physicians handle the flow of information from medical history, vital signs, genetic background, and lab testing into diagnosis and treatment

—funding Centers for Medicare and Medicaid Services and the National Institutes of Health projects demonstrating the value of innovation in health care

It is clear that our health care system needs better interoperability, improved data sharing, more balanced privacy controls, the development of rapid-learning feedback models, and deployment of predictive modeling for physician practices. Once these changes are put into place, the United States will be in a much stronger position to gain the benefits of personalized medicine for patients, caregivers, and the health system as a whole. The result will be a healthier population at a reasonable cost.

CHAPTER NINE

NEW DIGITAL MEDIA

A DECLINING ECONOMY HAS PLACED enormous fiscal pressures on news organizations around the globe. Simultaneous declines in ad revenues and circulation levels have undermined the traditional business model of newspapers, radio, and television. At the same time, the emergence of new competitors in the form of Internet websites, blogs, Facebook, YouTube, and Twitter has intensified the competition among media outlets.[1]

This set of circumstances brought financial disaster to the American media, and at an inopportune time. In the midst of economic problems, two wars, and major domestic policy challenges, an informed and engaged citizenry is of utmost importance.

Aspen Institute chief executive officer Walter Isaacson has written a cover story for *Time* magazine on the embattled state of print journalism, entitled "How to Save Your Newspaper." The title of Paul Starr's article in the *New Republic* suggests a dismal future: "Goodbye to the Age of Newspapers (Hello to a New Era of Corruption)." John Nichols and Robert W. McChesney have written a book titled *The Death and Life of American Journalism*.[2]

There is no question that the country is witnessing a fundamental economic and technological transformation in journalism. Traditional business models are dying, and new ones are still being developed. Increasingly, news consumers have shifted from a few general-purpose sources, such as the evening television news and local newspaper, to a larger number of niche publications and specialized news sources. Like diners at a smorgasbord, news seekers graze business news from one source, weather from another, sports, politics, and commentary from yet others.

Media attention to education has been especially hard hit. Even though most people recognize the importance of education for social equality, economic development, and long-term civic engagement, education issues garner little attention from the news media. When education is covered, it typically is in the context of personal scandals or school crime stories. There is little reporting on school reform, curricular innovations, or teacher performance.

In the multifaceted, new-media universe, we need an information strategy for the news industry that expands on the strengths of digital media, such as diversity, immediacy, and interactivity, while encouraging in-depth coverage. Several steps could be taken involving digital payments, news presentation, and changes in antitrust and tax laws to make sure that our new digital press achieves its full potential for our information ecosystem.

THE ELEMENTS OF FINANCIAL DISASTER

Virtually every trend in traditional print and broadcast media consumption and in ad revenues is down and not likely to improve any time soon. Daily print circulation has declined from 62 million to 49 million nationwide over the past twenty years. Circulation for leading newspapers such as the *New York Times* is down more than 10 percent.[3]

With an economy in recession, advertising revenue at many papers has dropped by 25 percent since 2006. Newspapers such as the *Rocky Mountain News* have closed for good, and the *Seattle Post-Intelligencer* has become online-only. Others, such as the *Chicago Tribune, Minneapolis Star Tribune,* and *Philadelphia*

Inquirer, went into bankruptcy. The *Ann Arbor News* has stopped printing and moved online.

Double-digit layoffs are occurring at many media outlets. The news staff at the *Los Angeles Times* has gone from 1,200 to 600 in the past decade. In a bid to cut costs, foreign bureaus have been closed, as have many D.C. news bureaus. Papers are cutting book reviews, film sections, and magazines.

At the same time, the popularity of the Internet has led many people to forsake traditional print and broadcast information for free online content. Nielsen Online estimates that 75 million Americans read papers online. For the first time in history, that number is higher than the comparable figure for print sources. Alone among traditional media, audience numbers for National Public Radio rose by 9 percent from 2007 to 2008.

The convergence of social, economic, and technological trends has been disastrous for print newspapers. The global recession has unfolded at the very time that old-media outlets face their most serious competition. Yet a new-media universe is emerging that offers several virtues from the standpoint of citizen involvement, democratic accountability, and civic discourse.

PROBLEMATIC FEATURES OF THE CURRENT MEDIA ERA

Recent economic woes have challenged media organizations in their mission to inform public discourse. "News is a business of substance and providing information for the sake of democracy," Bill Buzenberg, the executive director of the Center for Public Integrity, has said. Every theory of democracy requires news organizations that inform people about current events and help citizens hold leaders accountable for their actions.

In the past, subscriptions and advertising provided sufficient revenues to cover both basic news—weather, sports, business, and obituaries—and public policy and political events, including investigative journalism. This general-purpose role of news outlets allowed them to cover a variety of topics without having to charge consumers for every specific item in the paper.

That business model has collapsed. News organizations have lost subscribers and advertisers and are facing nimble new web-based competitors. These web outlets typically are small, specialized, and have far lower personnel and overhead costs than mainstream print publishers. In a fragmented news environment that includes a wide variety of platforms and delivery systems, it is harder to finance, for example, high-quality coverage of education. Thus did an earlier report find that education stories in national outlets constituted only 1.4 percent of the total front page and prime news hour coverage during the first nine months of 2009.[4]

But it is not just the quantity of news coverage that is problematic. That same report also notes that a disproportionate amount of local and national coverage focuses on the politics as opposed to actual policy discussions. Local news outlets did a better job in these matters, but they too face economic difficulties.

With declining time, money, and expertise, there are fewer incentives and outlets for longer analytical pieces. Dale Mezzacappa, president of the Education Writers Association, notes that

> you can get reporters who have developed a pretty good expertise of how to observe a classroom and know good research from bad research and everything, but they don't have the time the way newspapers are structured today to actually spend what's needed to really do in-depth stories that advance the issues and inform the public, and enlighten the public. They always get pulled off on the daily stories, which are not unimportant, but which prevent the longer-term projects.

REIMAGINING THE NEWS MEDIA

An economic crisis is a perfect situation in which to consider alternative models because news organizations have to think about different ways of doing business to stay alive. Nostalgia about the good old days of American journalism is misconceived. We are not going back to the Walter Cronkite era of a few television networks, wire services, and prestigious newspapers. The emerging news industry is large, edgy, noisy, and opinionated.

Although many bemoan the loss of responsibility and restraint in the new digital press, this system has a number of discernible benefits. It is more representative of America, more democratic, more engaging, and more diverse than its traditional counterparts. Unlike the so-called objective press of the twentieth century that largely ignored alternative lifestyles, racial minorities, and social and political views outside the mainstream, today we have the greatest diversity of viewpoints in American media history.[5] Ten thousand flowers have bloomed in the form of niche journalism, portals, news aggregators, citizen journalism, new content providers, and new vehicles for investigative journalism, and we need to figure out how to harness the positive elements of the digital press for a brighter future.

Niche Journalism

The Pew Research Center's Project for Excellence in Journalism released a 2010 report that documents the commonly held view that general-purpose news outlets are in serious and probably irreversible decline. For example, ABC television's Washington, D.C., staff dropped from forty-six people in 1985 to fifteen in 2008. NBC cut its Washington bureau from thirty-four to twenty during this period, and CBS went from thirty to sixteen staffers.

In the print arena, newspapers with Washington bureaus went from more than six hundred in 1985 to three hundred in 2008. Newhouse Newspapers, Copley News Service, the *Los Angeles Times*, the *Chicago Tribune*, the *Baltimore Sun,* and the *Hartford Courant* have closed their Washington bureaus. The *Washington Post* ended its separate business section and merged financial content with its front-page section. Some newspapers such as the

Christian Science Monitor have given up on print and migrated entirely to the web.

What received less attention in this report, though, was that niche and specialty publications are flourishing. For example, the number of Washington-based newsletters rose more than 50 percent during the past two decades, from 140 to 225. Specialty news outlets such as *Government Technology*, the *Energy Trader*, and *Food Chemical News* have proved popular with specialists in relevant fields. *Bloomberg News* is one of the most successful specialty outlets, with 275,000 clients paying a large annual fee for detailed business-related coverage. In public affairs, *Roll Call*, the *Hill*, the *National Journal*, the *Congressional Quarterly*, the *Huffington Post,* and Politico.com have carved out followings through their detailed reporting of political life inside the Beltway.

Commentators like to focus on the first part of that story while ignoring the second. Traditional outlets are giving way, but they are not leaving an empty vacuum. Rather, these sources are being replaced by new variants. A rich array of newsletters, specialty publications, digital newspapers, and websites provides detailed substantive information.

Portals and News Aggregators

Other media outlets have moved from being original content pro- viders to becoming news aggregators that serve large masses of readers and viewers. They synthesize existing coverage from a variety of sources and serve as a clearinghouse for news informa- tion. This includes portals such as Yahoo, Google News, AOL, and MSN, each of which attracts a large number of visitors every day. AOL announced recently that it planned to start reporting the news, rather than just serve as a news aggregator. For editor of its reporting division, AOL hired Melinda Henneberger, who has previously worked at *Newsweek*, the *New York Times,* and *Slate*.

The website RealClearPolitics.com is a news aggregator that republishes commentary and news analysis from newspapers and media outlets across the country. On its homepage, it links to columns by liberals, moderates, conservatives, and libertarians. As Cass Sunstein points out in his book *Republic.com*, this is exactly the way a fragmented media system serves the public interest, by providing a rich diversity of perspectives.[6]

In major policy areas, we see the emergence of policy-specific portals. In the area of education, for example, EdNews.org represents a leading source of online news about elementary, secondary, and postsecondary education issues. It reaches 1.3 million unique readers a month. ModernHealthCare.com has proved popular by providing the same type of coverage for the business and policy aspects of health care.

These mass and specialty aggregators offer helpful information for a democratic political system because they draw from a range of different content providers and gather the best material in one place. They serve a valuable function as go-to portals for news about particular topics.

Citizen Journalism and the Democratization of News Gathering

One of the most noteworthy features of the new-media system is the democratization of news gathering. In the old regime, professional journalists served as gatekeepers. They gathered the news, placed information in context, and decided what was important. However, present-day news gathering is more democratic because it involves a broader range of people serving as news gatherers, commentators, and interpreters of political events.

Citizen journalism takes a variety of different forms: instant news reporting from ordinary citizens, crisis coverage from eyewitnesses on the scene, and blogging and commentary. CNN encourages iReporters to upload video reporting, commentary, or analysis to the cable news network. Meanwhile, Current TV specializes in viewer-provided content, which is news or entertainment features from viewers around the world.

Bloggers provide commentary on virtually every topic. For example, by some estimates there are more than 5,000 blogs in the United States devoted to education. Richard Lee Colvin of Columbia University's Hechinger Institute on Education and the Media has a weblog outlining his views regarding problems of press coverage of education. There are blogs for current college students, such as *College v2* and *Year One*. Popular blogs include *Eduwonk* and *This Week in Education*. Activists seeking to reform schools have blogs called *Change Agency*, *D-Ed Reckoning*, *Education Intelligence Agency*, *Practical Theory*, and *Schools Matter*. There are blogs that focus on learning, such as *2 Cents Worth*, the *Informal Learning Blog*, and a *Random Walk in Learning*. Research-based blogs include *Free Range Librarian*, *Research Buzz*, *Deep Thinking*, and *Dissertation Research*. Teaching is the focus of *Are We Doing Anything Today?*, *Bud the Teacher*, NYC *Educator*, and *Teachers Teaching Teachers*. Instructional technology is the focus of *Bionic Teaching* and *Ed-Tech Avenue*.

These citizen-run outlets make news gathering more interactive and democratic than what previously existed. During times of crises, disasters, and emergencies, when professional journalists are not yet on the scene, new technology allows eyewitnesses to inform the rest of us. They provide timely and relevant information that supplements the work of trained news gatherers.

New Policy Content Providers

As traditional news organizations have cut budgets and laid off employees, policy content increasingly is coming from universities, think tanks, nonprofits, and foundations. Geneva Overholser, director of the School of Journalism at the University of Southern California's Annenberg School for Communication, has called for a new business model based on nonprofit news. The idea is that nonprofits and other organizations can step in and substitute where coverage begins to disappear in important areas.

The Pew Research Center estimates that in the past four years, $141 million of nonprofit funding has gone into new media outlets, including public broadcasting. However, this represents less than one-tenth of the $1.6 billion in newspaper resources lost during this time period.[7]

Researchers from a variety of knowledge industries disseminate findings to maximize impact and influence the public debate. Think tanks such as Brookings, Cato, Hoover, Heritage, the Center for American Progress, and the American Enterprise Institute release policy reports, analysis, commentary, and video questions and answers. Universities offer a rich supply of information on education in the form of books, articles, reports, and events. Many colleges and universities have expanded their public outreach activities, constructed television studios, and encouraged faculty members to explain their research to audiences beyond academia.

There are also new, community-based nonprofit newspapers supported by local foundations and individual donors. Chicago, Minneapolis, San Francisco, Seattle, and San Diego have witnessed the emergence of nonprofit papers seeking to fill the local news void created by the collapse of major metropolitan newspapers. San Francisco's Bay Area News Project was given $5 million by philanthropist Warren Hellman. The project features collaboration with KQED-FM and the University of California at Berkeley's Graduate School of Journalism. Students at the school and staff at the radio station work together to report on news of interest to the Bay area.[8]

To deepen policy content, foundations subsidize reporting and beat coverage in specific areas. The Spencer Foundation subsidizes reporting on education research at *Education Week*. In the health care area, the Kaiser Family Foundation has created a digital Kaiser Health News Service that undertakes original reporting on health care and makes that information available to news organizations free of charge. The Bill and Melinda Gates Foundation and the Carnegie Corporation provide expertise and financial support for National Public Radio to generate education-related reporting.

New Vehicles for Investigative Journalism

The biggest challenge for contemporary news organizations is investigative journalism. This domain requires more time and financial resources than daily coverage, commentary, blogging, or opinion writing. Most of the time, citizen journalists cannot be relied on to unearth government scandals or unethical behavior (though sometimes bloggers have broken major stories). Investigative stories take weeks and months to prepare and require substantial financial support for staff, research, and communications.

The Center for Public Integrity represents a new model for investigative journalism. This nonprofit undertakes investigative reports in a variety of areas and makes that information available free of charge to news organizations. The *Huffington Post* has set aside $1.75 million in conjunction with the Atlantic Philanthropies foundation to support ten investigative reporters who will write about the economy. The nonprofit ProPublica.org practices investigative journalism on leading policy issues. In a world of declining general-purpose news providers, these specialty outlets represent a way to encourage democratic accountability.

THE CASE OF EDUCATION COVERAGE

Concerned about declining coverage of education, a number of foundations provide grants to support reporting in that vital area. As budget cutbacks have slammed newsrooms and drops in revenue have led to staff reductions, this option has helped some outlets provide high-quality content.

This approach has been tried in other policy areas. For example, the Kaiser Family Foundation, which specializes in health care, has created a website that undertakes original reporting on health care and makes that information available to news organizations free of charge. In the area of investigative journalism, the

Center for Public Integrity and ProPublica.org (the recent winner of a Pulitzer Prize) practice investigative reporting in a variety of areas and make that information available to news organizations.

The Center for Public Integrity's Buzenberg thinks that non-profit news is adapting better than commercial news to the changing landscape of the industry. At the many conferences he has attended on the future of the news industry, he says it is the nonprofit sector that is "talking about technology and moving forward," in contrast with the old private sector models caught in the financial crisis.

The Gates Foundation and the Carnegie Corporation provide expertise and financial support for National Public Radio (NPR) to boost education-related reporting. The Gates Foundation has specifically funded education coverage on public radio, says Marie Groark, NPR's senior program officer, because it believes radio has an audience that "is stable, perhaps even growing," compared with other media outlets. It also sees the public radio audience as including opinion leaders whom foundation officials want to reach. Chris Williams, Gates Foundation program officer, says that the foundation wants to "increase the bandwidth, quality, and depth" of reporting and that foundations have a "critical role" to play as a supporter of education coverage.

Foundations such as Spencer, Gates, Carnegie, Hewlett, Joyce, Pew, Mott, MetLife, and Wallace support reporting at *Education Week*. That publication started as a nonprofit print outlet; its first issue appeared on September 7, 1981. Part of Editorial Projects in Education, it features a number of different components such as TeacherMagazine.org, TopSchoolJobs.org, and Digital Directions. *Education Week* was one of the first education publications to go online in 1995, and it has invested heavily since in its EdWeek.org website. The website has about 50,000 paying subscribers, and an estimated 260,000 see each issue because there are multiple readers for many subscriptions. Currently, 40 percent of its revenue comes from nonprint sources, according to its president and editor, Virginia Edwards.[9] Such support is helpful, Edwards says, when it covers "big broad swaths" of the field, ranging from teaching and district-based reform to the impact of the economic stimulus package. Funding for broad topics preserves editorial independence and allows writers and editors to decide which specific topics get covered. Many foundations do not underwrite general operating expenses but support topics linked to areas of coverage.

But foundation money does not entirely fill the gap between revenue and expenses. Nonprofits, no less than for-profit news outlets, have had to be entrepreneurial in seeking alternative income sources. *Education Week*'s webinars (online seminars) have proved popular, and the publication undertakes three of these each month, tripling last year's rate. They are able to draw 2,000 to 5,000 paying live watchers, plus another 2,000 to 5,000 people who view it later. The organization has sponsors for its webinars and attracts attendees from schools, government, and business.

Education Week's daily e-newsletter has 240,000 readers and regularly sells out its ad space. It offers Really Simple Syndication (RSS) e-mail feeds on seventy-five different topics, allowing recipients to personalize their news delivery. From time to time, it sponsors live forums in cities around the country to extend its brand and generate revenue and sponsorships. It is developing a book club for teachers and has content accessible through mobile devices.

Partnerships that further spread the publication's content to other people also gain access to new content. Soon, *Education Week* stories will appear on the Associated Press news wire. It also has partnerships with the McClatchey Wire and with education aggregators such as the Association for Supervision and Curriculum Development (ASCD) Smart Brief that bundles K–12 content from hundreds of sources. This allows writers to reach new audiences and develop alternative distribution systems.

Of course, all of these products complicate the lives of sales departments. Edwards notes that "sales used to sell one thing, print advertising. Now, . . . we have distinct products that we are selling that come in different flavors, such as live events, e-newsletters, webinars, and chats. Sales people have to be agile."

In making the transition from print to digital publication, education outlets have to navigate a delicate balance: between offering free online content to boost traffic and attract advertisers, on one hand, and continuing to attract paid subscribers to their print products and subscription parts of their web sites, on the other. *Education Week* carefully monitors this balance. So far, it has found that only half of its print subscribers have registered to receive free access to that part of the website requiring payment. Getting users to access online material for which they have already paid through print subscriptions represents a key goal of the new digital world. If people do not register, it becomes harder to move toward online payment systems for premium content.

Some worry about the long-term sustainability of a nonprofit approach since it is at least partly dependent on foundations or individual donors supporting education journalism. Funding priorities may change, or funders and grantees may come to disagree with editorial policy on important issues. According to Doug Lederman, an editor at the for-profit InsideHigherEd.com, "The foundation model is designed to invest, build, and get out." It is a temporary answer, but not sustainable in the long run.

Foundations do not have unlimited resources for program support. Chris Williams of the Gates Foundation cautioned against thinking that foundations and nonprofits could "offset costs and take over issues" from news organizations, because of the uncertainty of future budgets. In fact, he describes such a scenario as "the worst thing" that could happen, because "who knows where foundation and nonprofit news funding will be in the future?"

The Carnegie Corporation does not have a large journalism program to support funding, so it focuses its resources on university journalism programs, *Education Week*, and National Public Radio, among other areas. A few years ago, Carnegie provided

NPR with funding for the express purpose of strengthening its local education coverage and conducting local training. In that grant, NPR specifically stated that it wanted to develop better rapport with individual school systems. Carnegie's vice president Susan King thinks that her foundation has a role in supporting NPR because it covers K–12 education (which is a Carnegie focus area) and reaches Carnegie's target audience of practitioners and policymakers.

Although the number of groups pursuing nonprofit journalism has increased, King says that the available funding has actually decreased. "Every time I turn around, there are a few more news organizations," she says, "but we don't have more money. We have less money." This is especially the case when all the groups are "competing for the same dollars." She says that some have suggested that foundations set up a consortium of sorts, whereby everyone would contribute to a common pool of funding to support nonprofit journalism.

This idea has not yet found wide support, but Buzenberg sees pooling resources as the key to sustaining nonprofit news over the long run. "Economies of scale and linking and interconnecting are important," he says. He believes that because nonprofit news is not about one company's ego and brand, the nonprofit news industry is better placed to use its collective resources efficiently and effectively. The possibilities of such partnerships include providing back offices for smaller investigative groups, collaborating on projects, or even doing joint fundraising. Building on that idea, the Center for Public Integrity partnered with two dozen news organizations to create a new Investigative News Network for watchdog journalism. Buzenberg notes that the Associated Press and National Public Radio both began as groups of newspapers or news stations that created an alliance. "It's the linking world we're in," he says, "and nonprofits can be more successful in adapting to that."

Of all those interviewed, Buzenberg is the most optimistic about the future of nonprofits news. He believes that the nonprofit sector is creating "a new ecosystem for information." It will be built on enthusiasm for the work, bolstered by the large stream of talent flowing from commercial news into the nonprofit world. He is cautiously optimistic about whether nonprofits have the necessary resources to compete in the news business. While acknowledging that the financial challenges are substantial, he notes that more and more models have been created that allow for individual support. "While only 16 percent of any audience will ever give money," he says, "that 16 percent plus institutional support can sustain a nonprofit news organization." NPR, he notes, has four times as many bureaus worldwide as any corporate network.

Buzenberg thinks that if a foundation were willing to invest the resources, the Kaiser Health News model could work for education, though he does not think that a one-size-fits-all news model could succeed. "Eventually, the best Internet models will likely be adopted and spread, but we're just not there yet," he says. "We're in a period of massive experimentation on many, many levels. . . . But there's a lot of hopefulness."

A number of for-profit print outlets have migrated to the web and support their coverage with paid subscriptions in conjunction with advertisements. The *Chronicle of Higher Education* is one such entity. Originally a print outlet, it now has a website, www. Chronicle.com, that reaches more than a million unique visitors each month. The *Chronicle* has seventy full-time editors and writers and seventeen foreign correspondents around the globe.

The first newspaper on the Internet, through a Gopher service in 1993, the *Chronicle* has a robust website. It is updated every day and sometimes on an hourly basis, depending on news flows. Its two e-mail news alerts a day are very popular. The morning e-mail is sent to 100,000 users, and there is another e-mail each evening, summarizing the day's events. The print edition has 70,000 subscribers, and there are web versions that are customized for mobile devices and smart phones.

Like *Education Week,* the *Chronicle* does not post all its content online free of charge. A subscription is required for what *Chronicle* president Phil Semas describes as its "best stuff," that is, material that is not easily accessed elsewhere. Since publication involves not just content but also a community and a conversation, Chronicle.com has blogs; it also allows readers to comment on articles and on forums about jobs. Sometimes, online conversations develop spontaneously without being initiated or led by the staff. This occurred, for example, after Hurricane Katrina struck New Orleans. While some blogs are written by outsiders, several staff reporters write blogs on hot topics such as athletics and college admissions.

The *Chronicle* plans to expand globally and will soon launch an edition of its website aimed at overseas audiences. "We need to become a global enterprise and cover more international issues," Semas says. International companies have a need to recruit Americans for jobs, and there is extensive interest outside the United States in what is happening in American higher education.

While growth clearly is moving to the Internet, Semas believes "[that] print still has a big role to play and that the print product is very robust. It will be around for some time." However, advertisers are presently more interested in the web than in print, and job services are moving toward the web.

Other outlets have been started with venture capital, and they rely mostly on advertising. The online InsideHigherEd.com was founded in 2004 by three people (Kathleen Collins, Scott Jaschik, and Doug Lederman) who previously worked for the *Chronicle of Higher Education.* Jaschik and Lederman quit the *Chronicle* on the same day in 2003. According to the *New York Times,* "There was speculation among the remaining staff that they had been forced out over differences with Mr. [Corbin] Gwaltney [the owner]," Lederman remarked. "We had come to the conclusion that we and the *Chronicle*'s owner had different visions about what the publication and company should be." Scott McLemee, who currently works at *Inside Higher Ed* as an "essayist at large," said, "The *Chronicle* was traditionally oriented towards the administration—there was a brief period where it tried to reach a larger constituency, and then it retrenched. . . . They have no particular interest in reaching anyone else."[10]

Inside Higher Ed sought to distinguish itself from other education publications by being online-only and making all of its content freely available. The founders told the *Washington Post*, "The *Chronicle*'s $82.50-a-year subscription price was out of reach for many in higher education, especially graduate students."[11] Including the three founders, *Inside Higher Ed* has a staff of nineteen writers. The site includes listings of available jobs, recent job change notices, and scheduled events and conferences. There are also news stories, regular opinion columns, blogs, and recordings of conferences on higher education topics.

In a recent interview, Jaschik said the goal of *Inside Higher Ed* was to "build a bigger audience" and "monetize it." "Colleges want to recruit and companies want to reach a college audience." Since its content is entirely free and online, its business model centers on advertising and job searches. Traditionally, colleges and universities advertised only a small percentage of their available jobs because of the cost of an ad. For example, they tended to advertise deans' positions but not junior admissions officers.

InsideHigherEd.com aims to expand the number of college jobs that are advertised nationally. It scales its job posting price to school enrollment, and there is no word or space limitation. Job applicants can link to in-depth information about the school, and some of its news articles post links to relevant positions. One of the unique features of contemporary job searches in higher education, according to Jaschik and fellow editor Doug Lederman, is spousal hires. Academic searches increasingly involve dual-career families. The *Inside Higher Ed* site has the ability to search for job titles in different subjects and focus on a particular area, such as greater Philadelphia or Boston. This allows applicants to look for joint positions in specific metropolitan areas across specific schools and particular disciplines. That has proved popular in faculty and administrative searches.

The site also focuses on corporate advertisers interested in academic audiences. These include technology companies such as Google, financial services firms such as TIAA-CREF and Fidelity, and college publishers. Book publishers, for example, might place the first chapter online and provide a link to interested readers where they can buy the rest of the book for a fee. A web publication, Jaschik explains, is "not just a print version"; it needs to be used more creatively to link to relevant information and be searchable with a high degree of precision.

InsideHigherEd.com runs 4,900 ads. The site attracts 600,000 unique monthly visitors, 10 percent of whom log on from abroad. Its daily e-mail news alert goes to 80,000 people. The site has found that e-mail alerts drive traffic each day and encourage people to look at stories of greatest interest to them. It also uses RSS feeds to customize content of interest to particular readers.

Like the *Chronicle, Inside Higher Ed* views blogs as a way to build community and create connections with readers. Unlike the *Chronicle,* it relies on outside writers, mostly in academia, for its blogs, not its own staff. And just as *Education Week* is building a partnership with Associated Press, *Inside Higher Ed* has a relationship with *USA Today* for the latter to publish one of its stories each day.

Jaschik sees the for-profit model as more sustainable over the long run than a foundation or nonprofit-funded approach. "We have hundreds of advertisers, not one, and no one departure would harm us." In contrast, he points out, a foundation model "is supported by a small number of donors, and this makes them vulnerable to their supporters."

Even in the case of investigative journalism, one area in which nonprofits and foundations have been quite active, Jaschik is skeptical. He says funders employ the "parachute approach," wanting a big splash. The problem, he notes, is that "the best big project is born out of beat reporting." It takes good contacts and top sources to break open scandals and exposés.

"The long-term solution," Jaschik says, "is the evolution of a bunch of business models. Our business model works because of our audience." InsideHigherEd.com is effective because it has a defined niche (colleges and universities) and a national scope. Most of the successful online publications have a clear niche and a well-defined audience. Jaschik predicts that successful news outlets will be online, have subject areas that are national in scope, be part of a community, and avoid talking down to readers. The niche has to be specific enough that people will pay for content.

AN INFORMATION STRATEGY FOR THE NEWS INDUSTRY

Through online news sites, citizen journalism, niche journalism, news aggregators, and new forms of policy content and investigative reporting, we now have a media system that looks very different from twentieth-century journalism. In that earlier period, all these functions were centralized in prestigious newspapers and broadcast networks. They were original content providers that delivered substantive content and investigated politicians.

Today, these functions are spread over a variety of news outlets. Specialty publications provide detailed substance. New content sources such as think tanks offer in-depth research. Nonprofits investigate government actions and raise questions about unsatisfactory behavior. Citizens stand ready with video cameras and impromptu questions to hold leaders accountable.

New digital technologies have facilitated a more democratic and participatory news industry than ever before. News gathering has become interactive and draws on a broad range of amateurs and professionals. Various parts of the system provide policy content and investigative journalism.

As long as consumers rely on multiple news outlets, there is little to fear in the noisy cacophony of contemporary voices. Readers understand that each outlet has a particular perspective and that they need to take what's offered with a grain of salt. Individual outlets can be loud and boisterous, but news consumption in the system as a whole is protected by the large number of voices represented.

There is little doubt that the new world of journalism is going to be digital and interactive. While print outlets clearly are going to be part of the media universe, all of the large education newspaper outlets have invested heavily in online platforms and see future growth coming largely through digital content. The old distinction between for-profit and nonprofit has broken down to some extent because virtually all outlets are experimenting with new revenue streams and supplementing standard education coverage with paid webinars, subscription events, advertising, book clubs, news alerts, RSS feeds, chats, and blogs.

In the end, it is impossible to predict which business models will be successful. As Semas notes, "Anyone who says they can predict [the future], I immediately mistrust them." But Google's chief executive officer, Eric Schmidt, recently told newspaper editors that the company's new business model would be based on "a combination of advertising and subscription revenue."[12] He said it was necessary "to experiment with everything from social media to personalized content to engage readers." The virtue of the current period, he said, is that "technology allows you to talk directly to your users" and that new delivery systems would emphasize smart phones, electronic readers, and mobile devices.[13]

The movement toward a new digital press is fundamentally shifting the nature of the news industry. Speaking at a 2010 American Educational Research Association panel about traditional print journalism based on paid subscriptions, Scott Jaschik concluded that "journalism is dead and a new version is taking place. It is a more inclusive model and doesn't privilege elites." Rather than engage in the "politics of nostalgia" and yearn for a recreation of the old-media order, he argued, executives must adapt to the new environment and create business models that make sense in a digital world.

Monetizing Content

The clear dilemma facing all media is how to get readers or advertisers to pay for online content. Determining how to migrate from an ecosystem with a large amount of free online material to one with paid content is the chief contemporary puzzle. Leading news providers need revenue to support the creation of original content. While news aggregators and social media outlets can synthesize content from various sources, news providers need to cover events and provide material for commentary, analysis, and republication.

Niche publications already have figured out how to get people to pay for online content. Outlets that have a well-defined niche and offer content that is hard to get elsewhere have been successful at developing premium content and subscription web sites. If users need stock market information, economic data, or sports coverage, they will pay.

The challenge for areas that are not specialized or where content is available elsewhere is getting readers and advertisers to pay. Without a clear niche or specialized content, it is hard to generate the revenues necessary to sustain a vibrant news organization.

Time will tell whether general-purpose newspapers can persuade people to pay for content. Along with the *Financial Times* and the *Wall Street Journal,* the *New York Times* now charges for content and, beginning in 2011, invited visitors to its website to pay for digital content after viewing twenty articles per month.[14]

News Corporation, the owner of the *Wall Street Journal,* uses a metered system in which people pay a sliding fee depending on how many stories they read. Rob Grimshaw, the managing director of FT.com (*Financial Times'* online version), says it would take "4 billion page views per month in order to make $50 million in revenue per year." Right now, the *Financial Times* online generates 85 million page views a month in conjunction with its print edition.[15]

Entrepreneur Steven Brill believes people will purchase online content if transaction costs are low and if paying is easy. He and his partners, Gordon Crovitz and Leo Hindery Jr., have built a company called Journalism Online that offers readers a metered system with a "simple-to-use account and password for hundreds of publications." This account allows people to purchase "subscriptions, day passes or single articles with one click."[16]

The idea is similar to an iTunes model proposed by Walter Isaacson of the Aspen Institute. When newspapers set up free websites in the 1990s, readers got used to accessing information without cost. As online content proliferated, it became increasingly difficult to impose fees because of the great supply of free online news. Micropayments could come in the form of a "one-click" digital system that allows "impulse purchases of a newspaper, magazine, article, blog or video for a penny, nickel, dime or whatever the creator chooses to charge." This would allow news organizations to adapt to the digital era and monetize their content.[17]

Commentator Michael Kinsley has disputed the premise behind Isaacson's concept. In a *New York Times* op-ed column, "You Can't Sell News by the Slice," he argues that the funds raised through this system would be insufficient for any major newspaper. Using the *New York Times* as an example, he calculates that the paper sells a million copies a day. A charge of $2 a month for the entire paper would generate only $24 million a year, well below the $1 billion in annual revenue currently collected by the paper.[18]

Monetizing online readership is the central challenge for the new journalism. User fees, paid advertisements, novel revenue streams, and indirect public support may generate solid and reliable income, but the specific designs depend on the segment of the audience being targeted. The very digital technology that torpedoed traditional business models may lay the groundwork for new approaches to sustaining journalism in areas beyond education.

More-Efficient Online News Presentation

We need to figure out how to load and display online news sources more efficiently. As Google's Eric Schmidt has noted in a recent talk to the Newspaper Association of America, "The online experience can be thought of as terrible compared to what I view as this wonderful experience with magazines and newspapers."[19] It takes longer to read online newspapers than print publications because digital stories are divided into multiple pages, pop-up ads interfere with news reading, and slow broadband creates annoying waits while text, audio, and video load on the screen.

As newspapers and magazines shift to the Internet, they need to be configured in a way that makes for faster and more pleasant news consumption. Consumers require an online experience that is at least as good as the print counterpart. If their experience is not rewarding, readers will be unwilling to pay for online content.

New Antitrust Provisions and a National Press Pass

Current antitrust provisions discourage use of multiple news outlets. In a 2008 article in *National Journal*, "How to Save Newspapers—and Why," the journalist Jonathan Rauch suggested a more general variant of the iTunes model, which he calls PressPass.net. According to this formulation, reader subscriptions would provide access to hundreds of digital news sites. Rather than a subscription allowing readership of a single outlet, PressPass would allow access to the universe of major news sources for a simple fee. This would allow readers to scan a variety of outlets and become their own news aggregators.[20]

Under current law, of course, this type of collective pricing mechanism would most likely violate antitrust collusion rules. It would probably take an act of Congress to legalize this approach, and legislators would need to relax antitrust rules. The PressPass would allow newspapers to work together in a way that benefits news consumers. Like the iTunes model, this provision would allow visitors to draw on multiple outlets. It would take advantage of the diversity of perspectives available in the contemporary news universe and allow more varied news outlets to survive.

Tax Credits for News Subscriptions

Right now, people filing an itemized tax return can deduct legitimate information expenses related to their jobs. For business-people, that could include subscriptions to financial publications. Professionals can deduct trade publications in their area of expertise. Others may write off newsletters, magazines, or other information sources that are related to their work.

Of course, most people do not itemize, so they are not able to take advantage of these tax incentives. According to the Internal Revenue Service, only 0.3 percent of individual filers itemize their deductions. Most Americans do not avail themselves of tax deductions or are not able to justify deductions because expenses are not related to their work.

Tax laws should be changed to provide a credit for news subscriptions for any filer, regardless of whether that person itemizes deductions or the news product is related to work. John Nichols and Robert McChesney suggest a tax credit for the first $200 spent on daily newspapers. But they attach unnecessary limiting conditions: the paper has to publish at least five times a week, be a minimum of twenty-four-pages long each day, and have less than 50 percent of its space devoted to advertising.[21]

The problem with this proposal is that it privileges daily newspapers over all other content, such as magazines, weekly newspapers, and online content. It also caps the credit for news consumption at $200, which would not cover a subscription to a single daily newspaper. A larger tax credit would encourage news consumption from a variety of online news sources.

A larger tax credit for news subscriptions would contribute to an informed citizenry, raise our civic dialogues, and help citizens hold leaders accountable. It is a policy change that would be good for American democracy. Subscription deductibility should not be tied to the job expense argument. Whether or not a news subscription is related to one's job, consuming and producing news represents a collective good for society, politics, and civic discourse. French president Nicolas Sarkozy recognized this virtue when he recently promised his country's eighteen-year-olds a free annual newspaper subscription.

The current economic meltdown has exposed the limits of a laissez-faire approach to the news. We need a national strategy that thinks about both the supply and the demand of public affairs information. As Nichols and McChesney note, that government plays no role in the flow of information is a myth. Newspapers and magazines have long benefited from subsidized postage rates. Television stations enjoy virtually free broadcast licenses. Cable outlets have achieved near-monopoly status in many communities around the country.[22]

An expansion of tax credits would help news organizations move toward charging for online content because those expenses would be subsidized for any American subscribing to a news source. Economic incentives have worked in areas from financial investment to child care and education. They are an effective way to encourage the acquisition of information and will help news organizations move into the digital era.

Newspapers as Tax-Exempt, Nonprofit Organizations

Maryland senator Benjamin L. Cardin has introduced the News-paper Revitalization Act, which would allow newspapers to become tax-exempt, nonprofit organizations under the U.S. tax code. Similar to public broadcasting stations, this change would offer two advantages to struggling newspapers. It would allow donors to deduct charitable contributions in support of coverage or operations, and it would make tax-exempt advertising and subscription revenue that the newspaper generated. There would be no restrictions on news coverage or campaign reporting, but newspapers would no longer be able to endorse candidates for office.

The greatest virtue of this proposal is that it identifies the news industry as an important national priority and provides the same type of favorable tax treatment enjoyed by nonprofits and charities that work in the public interest. By offering important financial support, this legislation would inject much-needed assistance to an industry that is important to the long-run well-being of the country.

CHAPTER TEN

SAFEGUARDING PRIVACY AND SECURITY

PRIVACY AND SECURITY REPRESENT MAJOR concerns for the American public. According to survey data, a large number of Americans are concerned about the confidentiality of online information and the security of electronic transactions. For example, 62 percent of adults in a national poll felt that use of electronic records makes it more difficult to ensure personal privacy. In addition, 75 percent of Internet users worried about websites' sharing information without users' permission.[1]

On the surface, the advent of new technologies appears to increase questions regarding privacy and security, especially with government agencies dealing with sensitive information.[2] By its very nature, storing information or accessing services through remote providers would seem to raise the level of risk. Many people feel more confident about security on their own desktop and laptop computers since those machines are in their personal possession.

However, many service providers take security seriously and have safeguards in place that are at least as good as those available to desktop, laptop, or mobile computing. Generally, the weak link in security is the human factor more than the platform itself. As long as we learn the lessons of past breakdowns, new technologies such as cloud computing have the potential to generate innovation without sacrificing privacy and security.

CLOUD VERSUS DESKTOP VULNERABILITIES

Computer users perceive higher security on desktop and laptop computers because they physically control those devices, and lower security with the cloud because information is stored remotely through third-party commercial providers or public clouds featuring multitenancy. In reality, though, there are serious security threats to all electronic information, regardless of platform.[3]

Desktop, laptop, and mobile devices have security problems owing to theft, loss, or compromised information. One study has found that 637,000 laptops are lost or stolen each year in U.S. airports. About 53 percent of individuals who have lost laptops reported there was confidential company information on the laptop.[4] Indeed, the Federal Trade Commission says that the most common places for lost or stolen laptops are airports, hotels, and automobiles.

In addition, 8 million cell phones (including 700,000 smart phones) are lost or stolen each year. A number of these have sensitive personal or corporate information, especially as people conduct more of their lives through mobile devices.[5]

Since individuals are in charge of password protection on desktop or laptop devices, security is often weak. Individual consumers rarely upgrade with security patches as vulnerabilities are identified. They do not want to be bothered with "strong" passwords and instead use passwords linked to their pet, birthday, anniversary, education, or sports team. With the popularity of mobile devices that are often operated with little password or security protections, the risk of compromised information is high.

Although the cloud is often perceived as less secure because it is outside the physical control of the user, cloud service providers tend to employ much more rigorous security standards than providers to local computers.[6] Many use encrypted or highly secure passwords featuring complex combinations of letters, numbers, and characters. Google, for example, encrypts its Gmail application through https mode.[7] Most providers insist on high-security facilities having armed guards, limited access, and detailed employee background checks.

In addition, cloud computing enables stronger security by providing better economies of scale. The best security often is expensive and beyond the financial capacity of individual consumers or small businesses. It requires substantial investment and technical expertise up front. Cloud providers with a large clientele are in a stronger position to provide these resources than an information ecosystem based on individual software sales. Cloud providers are responsible for security and update their systems automatically as patches are made available.

MAXIMIZING PRIVACY, SECURITY, AND INNOVATION

Regardless of the storage platform, there is no question that strong privacy and security rules are required.[8] People fear adverse consequences from the release of confidential information. As Marjory Blumenthal points out, there are risks that the cloud could create a "platform for malice" that will be exploited as cloud computing rises in popularity.[9]

However, privacy rules must strike the right balance between privacy, security, and innovation. Too much emphasis on privacy means we lose the benefits of collaboration and information sharing. With many new applications arising that connect people and data sets, we need to find ways to maximize each of these three features. Too many of our computing systems cannot talk with one another, and this robs us of the ability to link datasets and perform transactions requiring connected information systems.

This is seen clearly in the health care area, where distinct data sets exist in terms of medical records, pharmaceutical orders, and administrative datasets. It is hard to integrate this information and learn what treatments work or do not work. But there are also connectivity problems in terms of national security, international commerce, and mass entertainment. Tremendous innovation is taking place in each of these areas, but interoperability problems have been difficult to address.

Cloud computing represents a way to deal with system integration and connectivity. Internet-based applications help us overcome problems of portability and transferability. By integrating information through cloud storage and services, it is easier to link information and transfer material across different types of computer systems. Many cloud systems are designed for information sharing. Since users have direct control over access to particular kinds of data, they can enable data sharing while still retaining control over information.

To move forward with cloud applications in areas such as health care, national security, and government efficiency, we need to address issues of access control, transparency and disclosure, security safeguards, and international differences. Specific actions in each of these areas would enable us to achieve the full advantages of technology innovation.

DIFFERENT TYPES OF THREATS

Moving information to a cloud introduces new classes of vulnerabilities in and of itself.[10] It creates new surfaces to attack and new interfaces to exploit. When those network resources are built on systems, platforms, and applications shared with others, problems are introduced. The control mechanism itself can be attacked, breaking down isolation between users and potentially allowing another user to access data or resources.

Some threats have different dynamics over the Internet. In a classic IT architecture, personal computers may be at risk of compromise through attacks exploiting local applications such as browsers or document viewers. If less data are stored locally, less are immediately at risk, but now the attacker could compromise credentials to gain access to the user's cloud privileges.

All organizations are vulnerable to attack from a trusted insider, but moving things to the cloud can raise the costs of misplaced trust. A cloud system with a well-thought-out identity interface and a clear access control system can restrict access and foster accountability. However, a unified data system with more people accessing more different types of data through more applications can actually make it harder to appropriately limit access and detect misuse.[11]

In turning over control of some aspects of their information systems to a cloud provider, users face a classic intermediary problem. When you rely on an intermediary, you cannot be certain that the agent is acting completely on your best interest, particularly when the two interests (such as profits) diverge. How does the client know what the provider is doing? This is further complicated by the fact that security is not an absolute property, and developing metrics to compare the relative attributes of different systems and approaches has proved quite difficult.[12]

Providers cannot offer an absolute guarantee against a bad outcome, so they must convince clients that they have taken adequate precautions. This trust must be extended to third parties contracted by the provider, over which the original client has even less control and little to no information. However, a study of cloud agreement terms and conditions frequently disclaims any responsibility for data security.[13] Responsibility for security lies with the user, even in instances in which the user cannot be expected to take any constructive or defensive action.

ENHANCING ACCESS CONTROL

In safeguarding privacy and security, access control is a crucial protection.[14] In the contemporary period, protecting privacy is challenging because people voluntarily place information in the public sphere through social networking sites, blogs, the Internet, and electronic communications. This flood of quasi-public information elevates the importance of access control in information systems.

The virtue of technology in general is that it helps people make access decisions in more refined and differentiated ways. User control allows people to employ technology that enables users to determine who has access to what levels of data for what time period. The cloud enables a more nuanced level of access control.

Digital systems furthermore produce a level of accountability that is not possible with paper records. When there are unwanted intrusions, technology makes it easier to identify abusers and enforce penalties. Intruders leave digital fingerprints that allow authorities to determine who saw particular records, what files they looked at, and how long the intruders browsed various parts of the site.

National surveys have found that 65 percent of social networking users have adjusted their privacy settings to reflect their personal preferences.[15] Online advertisers have noticed that when users are given the opportunity to control their preferences, 67 percent view settings but do not change them, 26 percent alter their preference settings, and 7 percent opt out.[16]

Technology allows sites to develop new ways of involving users in decisionmaking. For example, some providers such as Facebook and AT&T have established public comment periods or online voting mechanisms on proposed changes. This enables users to make suggestions or decide on suggested alterations.

With cloud applications, it also is possible to digitally tag data and information in ways that allow information owners to designate who is allowed to see which parts of the stored material and then to track unwanted intruders. Each of these actions helps system managers control information and balance innovation, on the one hand, with privacy and security, on the other.

IMPROVING TRANSPARENCY AND DISCLOSURE

Transparency is key to consumer confidence. Providers need to improve disclosure of privacy notices and business practices so that users know what is being offered. Companies should provide clear and understandable privacy notices and notify consumers about privacy issues when data are collected, not after the fact.[17]

Increasingly, consumer advocates are calling for "layered" disclosure statements in which people are asked about their privacy rights at the time information is collected. This means that notices should be sent when the particular information identifying a specific person is gathered. The notice should outline why certain information is being collected and under what conditions it would be disclosed to others.

One of the virtues of the cloud is its potential for real-time performance data and on-demand transparency. For example, companies can offer real-time information on cloud downtime and data breaches. In the case of the latter, there can be fast consumer notification of security violations. On-demand transparency gives consumers instantaneous information. This is an advantage that was not available in the era of paper-based records or some types of desktop computing.

SAFEGUARDING SECURITY

There are a number of different Internet security approaches designed to maximize information confidentiality in terms of consumer interfaces with computing systems. Some cloud providers use encryption techniques to prevent unauthorized snooping. Other companies employ "two-factor authentication." Under this approach, the service provider sends a separate verification code to a phone or asks for a verification code self-generated by the user following login and entry of a password. Completion of those two steps allows a person to access information with greater certainty that he or she is the actual user of that system.

Others suggest authentication through the use of "open identity" systems that direct users during login to another site for authentication.[18] Rather than having to develop new user names and passwords for every individual site, this approach enables users to rely on a single site with secure authentication to gain access. This offers the virtue of simplifying the maze of passwords that afflict nearly every contemporary computer user.

In addition, cloud-based sites have the ability to notify people of unusual activity on their accounts or provide warnings about visitation to high-risk sites with a history of malware or phishing activity. If someone logs on from a geographic area distant from where the owner of the account lives, the site may generate additional security questions or protocols.

One of the gravest security risks of current computing systems is old browsers with known security gaps. Exploiting gaps in these browsers is easy to do, and some users have not downloaded security patches that would fix flaws in this software, leaving them doubly vulnerable to hacking.

Cloud applications help with this problem because Internet servers provide automatic updates to protect site integrity with new security patches installed as they are made available. Rather than being dependent on the individual consumer to install a security patch, cloud providers automatically undertake that action, freeing consumers of the need to update security definitions.

THE NEED FOR CONGRESSIONAL ACTION

Most privacy rules were written well before the Internet was formally established and online content became ubiquitous, so it is important that Congress take action to ensure reasonable privacy and security. In 1986, when Congress approved the Electronic Communications Privacy Act (ECPA), few people used e-mail, the Internet, social networking, cloud computing, or file-sharing websites.[19]

As a result, the law was vague or inconsistent in how it treated privacy provisions. For example, privacy advocates point out that electronic communications currently have stronger safeguards if stored on local than remote file servers. "The main thing that's broken about ECPA is that it penalizes you for using cloud computing," said Marc Zwillinger, a privacy law attorney.[20]

We need to place privacy on the same footing for all users. It makes little sense to have weaker standards on one platform than on another. Consumers and government decisionmakers expect the same level of protection whether they are accessing information on a desktop, laptop, mobile device, or cloud storage system. The platform does not affect their personal practices or expectations of privacy.

One change that would be desirable concerns the process by which law enforcement agents can obtain electronic information. Instead of using a prosecutor's subpoena, legislation should require a probable-cause search warrant that has been approved by a judge. This would provide greater safeguards in terms of online content, pictures, geolocation data, and e-mails.[21]

PROTECTING SECURITY ON GOVERNMENT CLOUD PLATFORMS

Security represents a major consideration in cloud migrations for government agencies.[22] Indeed, a Government Accountability Office study undertaken by Gregory Wilshusen found that "22 or 24 major federal agencies reported that they are either concerned or very concerned about the potential information security risks associated with cloud computing. Risks include dependence on the security practices and assurances of a vendor, and the sharing of computing resources."[23]

The Computer Fraud and Abuse Act of 1986 needs to be updated to clarify its definition of presumed losses through cloud computing. The law is vague on whether the maximum penalty of up to five years in jail and as much as a $250,000 fine for unwarranted intrusions applies to the cloud data center as a whole or to each individual and business account that is accessed. Ideally, penalties should apply to each cloud account that is compromised. Otherwise, the penalties for unwanted cloud intrusions are artificially low given the magnitude of the possible losses.[24]

The National Institute of Standards and Technology and the General Services Administration currently are developing security standards for low-, moderate-, and high-risk applications. Organizations that have highly sensitive or classified information obviously require greater safeguards, both in terms of monitoring and firewalls. The federal government currently has rolled out low-risk cloud solutions and soon will be doing the same thing for moderate-risk applications.

To ensure security, the General Services Administration standard for e-mail is that systems be stored in at least "two different data center facilities at two different and distant geographic locations."[25] Eventually, the federal government aims to meet high-risk security needs for agencies such as the Department of Defense. It plans to roll out these applications for software, platform, and infrastructure needs this year.

Government agencies need to develop safeguards appropriate to the mission of each organization. With concern over cybersecurity threats, there are pressures to increase security safeguards and maintain secure facilities. But officials need to be aware of the costs and benefits of enhanced security safeguards. It is clear that the greater the need for highly secure storage and applications, the higher the cost of the cloud and the lower the possible cost savings that may come from cloud migration. Agencies with high security needs generally require that information be stored in secure facilities within the continental United States and operated by individuals with high-level security clearances who have passed background checks.

THE IMPORTANCE OF JUDICIAL ACTION

Differential privacy protection has emerged only recently. Judges have decided that people who share information with third parties have a diminished expectation of privacy and therefore do not have the same Fourth Amendment guarantees against unreasonable search or seizure by government authorities. By definition, material stored on a cloud involves voluntarily sharing material with a third party. In the eyes of judges, that puts privacy on cloud platforms on a lower level than that associated with desk-top, laptop, or mobile devices. This means that when law enforcement agents request e-mails, Internet usage, or other electronic communications, they face a lower threshold for accessibility with cloud applications.

This approach to cloud privacy should be clarified to ensure that consumers have an accurate understanding of which platforms they are using and how privacy protections vary across domains. During the course of a day, people shift quickly from cloud to desktop to flash drives without thought to how confidentiality rules differ. When they shift from a desktop to the cloud, most consumers are unaware that their privacy rights drop precipitously.

An analysis of cloud contracts has shown that many have varying legal features.[26] Some place restrictions on provider liability in terms of data integrity, confidentiality, and access. Right now, the burden is placed on consumers to read terms carefully and make sure their rights are protected in specific contracts. It is not clear that consumers have the sophistication or legal knowledge to understand the fine points of these agreements. It is important that safeguards be put in place to protect consumers from misleading or obscure legal provisions.

PART FIVE

CONCLUSION

CHAPTER ELEVEN

FACILITATING INNOVATION

THERE ARE MANY POSSIBILITIES FOR using digital technology in the private and public sectors to further social and political innovation. Recent advances in information technology have lowered the costs of communication and reduced the barriers to innovation in many fields. Those benefits enable new applications in government, business, health care, and communications, among other areas.

But there are a number of factors that complicate innovation. Successful implementation depends on a combination of shifting organizational routines, creating proper incentives, and enacting policies that facilitate innovation. Each of these characteristics is vital for long-term effectiveness.

PROMOTING INNOVATION WITHIN ORGANIZATIONS

Leadership sets the overall tone and atmosphere for inventiveness. Businesses and government agencies that do well in innovation typically have leaders who value innovation, make it a high priority, and create conditions that facilitate new ways of thinking.

One of the biggest barriers to innovation is agency structures that isolate related functions and inhibit cross-organization cooperation. By its very nature, innovation requires an ability to think independently and across structural boundaries. For example,

185

much of current health care innovation involves connecting isolated databases and integrating knowledge management. Many inefficiencies arise from the tendency in many organizations to wall off functions and discourage agency integration.

Organizations need to overcome institutional fragmentation and disseminate best practices across the agency. Officials should learn from one another about what works and what does not. They should share information so that successes diffuse rapidly and failures get short-circuited. The best innovation occurs when structures are aligned so as to encourage new ways of thinking.

CREATING PROPER INCENTIVES

It is crucial to align organization incentives with individual innovation. Government agencies need to align their own incentive structures so that executives are rewarded for new ideas. They need to break down agency silos and build connectivity across agencies and government missions.

In many organizations, there needs to be increased funding for technology innovation. This is especially the case in the public sector. Money for government innovation trails that of the private sector. As noted in chapter 6, companies spend 2.5 percent of their budget on technology, compared with 1.9 percent in state government.

This is one of the reasons why government innovation lags that of private companies. Agencies simply do not have access to the same level of financial resources as the private sector. Money is not everything in the public sector. But resources make it possible to purchase faster equipment, undertake market research, and customize software to organizational needs.

At the same time, the public sector needs to become more efficient in its use of the technology dollars it has. One way it could do this would involve moving to government-wide contracts designed to improve economics of scale. In many countries, individual departments negotiate their own technology contracts.

This limits the government's ability to take advantage of economies of scale and get the efficiencies of large-scale contracts. Rather than using the large scale of the government to negotiate favorable deals, the agency-by-agency approach undermines the purchasing power of the public sector and limits its ability to derive the greatest value from its contracts and purchases.

Regardless of whether the organization is in the public or the private sector, there need to be consequences for good and bad performance. Future agency resources should be tied to performance metrics in transparency, accountability, productivity, and efficiency. This creates strong incentives for agencies to innovate. If officials believe their future budgets are linked to performance, they will focus carefully on those performance metrics.

RECOGNIZING DIFFERENCES BETWEEN INDIVIDUAL AND ORGANIZATIONAL INNOVATION

Analysts often think about technology adoption at the individual level. They examine cases in which consumers have purchased new gadgets or government agencies have adopted new delivery systems. Innovation takes place when more people purchase electronic devices or organizations computerize paper-based processes.

But technology innovation can take place in different forms. Adoption varies depending on whether the innovation is based on individual adoption or machine-to-machine interconnections. Getting individuals to innovate is different from building innovation through machines. The former requires people to make changes, whereas the latter requires getting organizations to change routines.

Increasingly in the future, technology innovation will take place when organizations connect computers and facilitate fundamental changes outside the context of individual consumer decisions. For example, smart energy grids operate through smart meters and smart appliances.[1] Computers connected to the grid monitor consumption patterns and provide data on peak versus off-peak usage costs. Smart appliances can turn on dishwashers and water heaters at set times to take advantage of off-peak pricing.

Machine-based innovation differs from individual adoption because it is driven by institutional decisionmaking. Once machine connections are set up and accepted by individual consumers, people do not have to make decisions for the innovation to unfold. In this type of innovation, automated processes drive change, more so than individual behavior.

One of the virtues of machine-based innovation is that adoption can advance without consumers' having to make decisions. As long as people do not object to the outcomes being pursued by the automated connections, technology innovation can grow by leaps and bounds. One of the strengths of this form of innovation is that it can produce tremendous advances in adoption rates in a short amount of time.

ENCOURAGING INNOVATION THROUGH POLICY CHANGES

A number of specific policy actions can speed adoption among individuals and organizations. These include moves to leverage private resources; extend the research and development tax credit; improve science, technology, engineering, and math training; commercialize university inventions; reform immigration policy; and strengthen the patent system.

Leveraging Private Resources

With the limited resources of public agencies, those who are serious about innovation should figure out how to use public funds for technology infrastructure to leverage private sector resources. Government departments do not have the money to finance as much technology innovation as is needed. In many areas, much of the successful innovation is going to be funded by private companies.

For example, the Federal Communications Commission has estimated that making high-speed broadband accessible to nearly all Americans would cost $350 billion.[2] Yet the federal government has budgeted only $7 billion for broadband infrastructure development. This means that 98 percent of the broadband costs will be borne by telecommunication and other private corporations.

In this financing model, government cannot dictate the course of development. Instead, it must work with businesses to leverage private dollars for the greatest impact. The most critical role of the government should be to encourage infrastructure development on the part of businesses and make sure those investments yield the greatest results.

Extending Research and Development Tax Credit

To expedite private investment, the national government needs to make permanent the research and development tax credit for private companies investing in innovation. Tax credits create incentives for businesses to invest and are vital to long-term innovation development. This is one of the most important ways the government can support the technology innovation of the private sector.

Improving Science, Technology, Engineering, and Math Training

The United States needs better training in science, technology, engineering, and math. Few American students are getting advanced graduate degrees in these fields, even though these areas are among the most important for next-generation innovation. Many of America's past accomplishments rest on discoveries and advances in these basic science fields.

Just as the country invested in science fields following the shock of the Russian launch of its Sputnik satellite in 1957, America needs to invest additional resources in basic sciences. This includes research and development money for higher education, improved support for graduate students seeking advanced degrees in science fields, and improved training of high school science and math teachers through the development of a master teacher corps.

Commercializing University Ideas

Universities are crucial for national innovation. Many of our most successful economic development areas in the United States have arisen around institutions of higher learning. For example, a number of Silicon Valley technology firms benefited from proximity to Stanford University and the University of California at Berkeley. Route 128 firms gained from the cluster of Harvard University, Massachusetts Institute of Technology, and other Boston area universities. The Research Triangle emerged with the help of the University of North Carolina, Duke University, and North Carolina State University.

Yet despite these successes, we must do more to improve the commercialization of university inventions. It is important for universities to streamline their processes and get a bigger bang for their buck. The federal government invests $147 billion in support of research and development. Of that, $90 billion goes to institutions of higher learning. This money generates only around $2.5 billion in licensing fees for universities.

Faculty incentives and university resource allocation need to be better aligned to encourage technology transfer and commercialization. University licensing offices should speed up their review processes, and chief technology officers should be recruited and compensated in ways that encourage and reward innovation. Universities should consider profiting from inventions not just through royalties and licensing fees but also through equity stakes in new companies. There should be greater transparency on federally funded grant activities in terms of how faculty members can maximize impact and commercialization.

Nearly all universities require faculty members and postdoctoral research fellows to operate through that school's technology licensing office. Professors typically must submit ideas for review and licensing and must share revenues with the university. Since discoveries are often made on university time or through the use of school equipment or personnel, universities negotiate licensing agreements with individual faculty members. This can be a time-consuming process because each discovery must be evaluated for its commercialization potential and specific contracts and agreements negotiated with each inventor.

Capital is a crucial ingredient in the commercialization of knowledge because financial resources are required to build prototypes, form companies, and take a product to market. Aside from family members and friends, capital can come from angel investors, venture capitalists, or corporations. Angel investors typically are involved at an early stage, while venture capitalists more often provide resources later in the project. Corporations invest money if the product is related to their own product line or they feel the invention can make them money.

Some universities have developed "proof-of-concept centers" that provide funding and commercialization expertise to early-stage innovators so that they can get advice on promising inventions.[3] The notion behind this idea is that if faculty members receive feedback early in the invention process, later delays will be reduced.

Express licenses have been proposed whereby universities employ standard licensing agreements for routine inventions.[4] Rather than undertaking independent reviews of each new idea and developing unique agreements for that particular product, universities can cut the time to development by streamlining licensing approval.

Still another idea is to provide faculty members with commercialization mentoring and coaching.[5] The thought is that improved guidance on how to file patents, license ideas, and attract needed capital would speed up the innovation cycle and reduce the time required for commercialization. Most scientists lack experience on business formation, investment plans, capital attraction, and marketing. Providing help on each of those fronts is likely to expedite commercialization activities.

Universities should provide help to faculty members and students on how to market products, incorporate companies, and attract venture capital. Many inventors have little background in commercializing their ideas and building businesses. Even if they wanted to spin off a business, they would not be sure how to do so. It would be helpful to them if schools offered how-to advice on technology transfer and creating a prototype of an idea.

Innovation prizes, incentive pay, faculty bonuses, and seed funding have been proposed that reward outstanding ideas. The view is that universities (or foundations) can create monetary award prizes that go to the best ideas. This would give inventors and institutions of higher education financial incentives beyond the market value of their ideas to commercialize their discoveries.

Sharon Belenzon and Mark Schankerman, for example, have found that incentive pay for technology licensing officials was associated with a 30 to 40 percent improvement in average university licensing fees.[6] Private universities were more likely to adopt incentive pay than public universities, and government licensing constraints reduced the creation of start-up firms and the amount of licensing revenue.

Reforming Immigration Policy

Immigration policy is important for innovation. In the years leading up to World War II, the United States recruited Europe's top talent for our nuclear program. Scientists such as Albert Einstein, Enrico Fermi, and Edward Teller immigrated to America and played an instrumental role in securing our country's future and developing its nuclear advantage.

Today, we need to rethink our immigration policy. We should value brains, talent, and special skills to attract more individuals with the potential to enhance American innovation and competitiveness, increasing the odds for economic prosperity down the road.

At a time of high unemployment, we need scientific innovators who start new businesses and create high-paying jobs. One study has found that 25 percent of all the technology and engineering businesses launched in the United States between 1995 and 2005 had a foreign-born founder. In Silicon Valley, that number was 52.4 percent. Much of the high-tech boom of recent years has rested on immigrant entrepreneurship.[7]

Right now, only 15 percent of our annual visas are set aside for employment purposes. Of these, only a small number of H-1B visas (65,000) are reserved for "specialty occupations" such as scientist, engineer, and technological expert.

The number reserved for scientists and engineers is drastically below the figure allowed between 1999 and 2004, when the U.S. federal government set aside up to 195,000 visas each year for H-1B entry. The idea was that scientific innovators were so important for long-term economic development that we needed to boost the number set aside for those specialty professions.

Other countries such as Canada, the United Kingdom, and Australia are more strategic in viewing immigration as a way to attract foreign talent. Canada, for example, explicitly targets foreign workers in categories of short supply who can contribute to its economy. It admits around 265,000 immigrants each year. Of these, 154,000 visas (58 percent) are set aside for economic purposes, such as skilled workers or live-in caregivers, whereas 71,000 (26 percent) are devoted to family reunification.

Canada's percentages are nearly the reverse of national policy in the United States. Here, nearly two-thirds of visas (64 percent) are devoted to family reunification by which American citizens of foreign birth can sponsor spouses, parents, children, aunts, and uncles for citizenship. Unlike other countries whose leaders understand the value of skilled and seasonal labor for long-term economic development, we continue to place a low priority on admitting immigrants with economic talents.

We need government policy that encourages positive public opinion. If we expect citizens to support immigration, the national benefits must be obvious. Today's policies focus public and media attention much more on immigrant costs than benefits. That explains why no one likes the status quo and comprehensive reform is difficult to enact.

Strengthening the Patent System

The U.S. patent system needs considerable improvement. Right now, there are routinely thirty-five-month delays in approving patent applications. This slows the movement of products to the marketplace and makes it difficult for inventors to protect their intellectual property.

The rules about what can be patented are unclear. Patent examiners hired by the U.S. Patent and Trademark Office make considerably less money than those working for law firms and private employers. So it is hard to find workers with the proper training and background to judge patents.

Internationally, some countries do not take protection of intellectual property very seriously. Microsoft chief executive officer Steven Ballmer told Chinese president Hu Jintao that only 10 percent of Chinese customers actually pay for the software they use. This example illustrates the need for stronger enforcement. Nonenforcement of existing rules robs inventors of the proceeds of their labors.

Setting Higher Performance Expectations

According to public opinion surveys, more than two-thirds of ordinary Americans believe the U.S. public sector is wasteful, is inefficient, and performs poorly. They do not feel that government officials serve the general public, and they think that little good is accomplished through governmental action.

In this situation, it is hard for Americans to believe the public sector can do better. When confronted with proposals for institutional reform or technological innovation, citizens' initial impulse is doubt regarding the efficacy of proposed changes. They typically do not think that government officials can do better than they are currently doing.

Although many will continue to believe that government is wasteful and inefficient, this book suggests that information technology can improve the effectiveness and efficiency of public sector performance. The extent of the actual savings depends on organizational improvement, willingness to reduce staff when cost economics require it, and the ability to deliver information and services more efficiently.

It will be a while before citizens become less cynical about government. The massive public mistrust that has developed over the past fifty years will not disappear overnight. But improved performance is a prerequisite for turning around public opinion. Citizens will never become more confident in government or society in general unless organizations improve their overall performance.

NOTES

CHAPTER ONE

1. This story comes from Andrew Glass, "Deal Thwarts Ban on Dial Telephones in Senate, June 29, 1930," Politico.com, June 29, 2010, p. 31.

2. Brian Arthur, *The Nature of Technology* (New York: Free Press, 2009).

3. This section draws on Martin Baily, Bruce Katz, and Darrell West, *Building a Long-Term Strategy for Growth through Innovation* (Brookings, 2011, forthcoming).

4. Christine Zhen-Wei Qiang, "Telecommunications and Economic Growth," World Bank, 2009, p. 1.

5. See Darrell West, "An International Look at High-Speed Broadband," Brookings, 2010.

6. Erik Brynjolfsson and Adam Saunders, *Wired for Innovation* (MIT Press, 2009).

7. Jane Fountain, *Building the Virtual State* (Brookings, 2001).

8. Clayton Christensen, *The Innovator's Dilemma* (New York: Harper, 2003).

9. This section draws on Darrell West and Jenny Lu, "Comparing Technology Innovation in the Private and Public Sectors," Brookings, 2009, pp. 1–3.

10. For information on the public sector, see Darrell West, *Digital Government: Technology and Public Sector Performance* (Princeton University Press, 2005), p. 4; for the private sector, see West and Lu, "Comparing Technology Innovation in the Private and Public Sectors."

CHAPTER TWO

1. CNN Politics, "Survey: Most Americans Believe Government Broken," February 21, 2010.

2. Gallup, "Trust in Government," 2009 (www.gallup.com/poll/5392/trust-government.aspx).

3. This chapter draws on Darrell West, "Public Sector Innovation: Making Government Faster, Smarter, and More Efficient," Brookings, June 2010.

4. Roger Yu, "Air Travelers Would Prefer to Do It Themselves," *USA Today,* October 11, 2010, p. 3B.

5. See Desmond Sheridan, *Innovation in the Biopharmaceutical Industry* (Hackensack, N.J.: World Scientific, 2007); and Lawton Burns, *The Business of Healthcare Innovation* (Cambridge University Press, 2005).

6. Martin Baily, Bruce Katz, and Darrell West, *Building a Long-Term Strategy for Growth through Innovation* (Brookings, 2011, forthcoming).

7. Darrell West, "State and Federal Electronic Government in the United States, 2008," Brookings, August 26, 2008.

8. Aaron Smith, "Government Online," Pew Internet and American Life Project, April 27, 2010.

9. Business Wire, "Online Transparency Drives E-Government Satisfaction, Trust in Government, According to Report from ForeSee Results," August 24, 2010.

10. Stephen Baker, "What's a Friend Worth?" *Business Week,* June 1, 2009, pp. 32–36.

11. Facebook Press Room, "Statistics" (www.facebook.com/press/info. php?factsheet#!/press/info.php?statistics); *Twitter Blog,* "Big Goals, Big Game, Big Records," June 18, 2010; comScore, "comScore Releases March 2010 U.S. Online Video Rankings," press release, April 29, 2010 (www. comscore.com/Press_Events/Press_Releases/2010/4/comScore_Releases_ March_2010_U.S._Online_Video_Ranking).

12. Beth Noveck, *Wiki Government: How Technology Can Make Government Better, Democracy Stronger, and Citizens More Powerful* (Brookings, 2009).

13. Erika Lovley, "Twitter as a Weapon," Politico.com, May 26, 2009.

14. Mark Drapeau, "Government 2.0: How Social Media Could Transform Gov PR," January 5, 2009 (www.pbs.org).

15. Adam Ostrow, "Does Congress Need Its Own Social Network?" *National Journal,* July 20, 2009; Congressional Management Foundation, "2007 Gold Mouse Report: Recognizing the Best Web Sites on Capitol Hill," Washington, 2007.

16. Thomas Mann and Norman Ornstein, *The Broken Branch* (Oxford University Press, 2008).

17. Michael Heaney, "Blogging Congress: Technological Change and the Politics of the Congressional Press Galleries," *PS,* April 2008, pp. 422–25; Patrick Dunleavy and others, "New Public Management Is Dead, Long Live Digital-Era Governance, *Journal of Public Administration Research and Theory* 16, no. 3 (2006): 467–94.

18. Thomas Patterson, *The Vanishing Voter* (New York: Knopf, 2002).

19. Matthew Baum and Tim Groeling, "New Media and the Polarization of American Political Discourse," *Political Communication* 25, no. 4 (2008): 1–21.

20. Daniel Carpenter, Kevin Esterling, and David Lazer, "Friends, Brokers, and Transitivity: Who Informs Whom in Washington Politics?" *Journal of Politics* 66, no. 1 (2004): 224–46.

21. Norman Ornstein, "Social Networking Sites May Be Model for the Government," *Roll Call,* July 15, 2009.

22. Ostrow, "Does Congress Need Its Own Social Network?"

23. Helen Margetts, Perri 6, and Christopher Hood, *Paradoxes of Modernization: Unintended Consequence of Public Policy Reform* (Oxford University Press, 2010).

CHAPTER THREE

1. This chapter draws on Darrell West, "Customer-Driven Medicine: How to Create a New Health Care System," Brookings, 2009.

2. Darrell West and Edward Miller, *Digital Medicine: Health Care in the Internet Era* (Brookings, 2009).

3. CNN Health, "Health Care Reach Expands with Wireless Monitoring," August 17, 2009.

4. World Health Organization, *Adherence to Long-Term Therapies: Evidence for Action,* 2003 (www.who.int/chp/knowledge/publications/adherence_full_report.pdf), pp. 7–10.

5. BBC News Technology, "Text Messages Prove a Life-Saver," January 29, 2003.

6. West, "Customer-Driven Medicine," p. 4.

7. West and Miller, *Digital Medicine*, p. 4.

8. Ibid., p. 1.

9. Harris Interactive Survey, "Many U.S. Adults Are Satisfied with Use of Their Personal Health Information," March 26, 2007 (www.Harris Interactive.com).

10. Ibid.

11. Sarah Arnquist, "Research Trove: Patients' Online Data," *New York Times,* August 24, 2009.

12. "Consumer Reports Rates Hospitals" (www.consumerreports.org/health/doctors-hospitals/hospital-ratings.htm).

13. Kathleen Parker, "Health Reform, Utah's Way," *Washington Post,* July 26, 2009.

14. David Leonhardt, "Fat Tax," *New York Times Magazine,* August 12, 2009.

15. World Health Organization, *World Health Statistics, 2010,* 2010 (www.who.int/whosis/whostat/EN_WHS10_Full.pdf), pp. 127–41.

16. Peter Neupert, "It's Not about Costs, It's about Enabling Transformation," *Washington Post,* July 21, 2009.

17. West and Miller, *Digital Medicine,* p. 1.

18. Robert Litan, "Vital Signs via Broadband: Remote Monitoring Technologies Transmit Savings," Better Health Care Together Coalition, October 24, 2008, p. 1.

19. PricewaterhouseCoopers Health Research Institute, "The Price of Excess: Identifying Waste in Healthcare Spending," 2008.

20. Helen Hughes Evans, "High Tech vs. 'High Touch': The Impact of Medical Technology on Patient Care," in *Sociomedical Perspectives on Patient*

Care, edited by J. M. Clair and R. M. Allman (University Press of Kentucky, 1993), pp. 83–95.

21. Edward Alan Miller, "Telemedicine and Doctor-Patient Communication," *Journal of Telemedicine and Telecare* 7, no. 1 (2001): 1–17.

CHAPTER FOUR

1. Ted Alford and Gwen Morton, "The Economics of Cloud Computing: Addressing the Benefits of Infrastructure in the Cloud," Booz Allen Hamilton, 2009, p. 1.

2. This chapter draws on Darrell West, "Saving Money through Cloud Computing," Brookings, 2010; and Darrell West, "Steps to Improve Cloud Computing in the Public Sector," Brookings, 2010.

3. Federal Communications Commission, *Connecting America: The National Broadband Plan,* 2010, p. 286.

4. Jeffrey Rayport and Andrew Heyward, "Envisioning the Cloud: The Next Computing Paradigm," MarketspaceNext Point of View, March 20, 2009, p. iii.

5. William Forrest, "Clearing the Air on Cloud Computing," McKinsey report, April, 2009, p. 5.

6. Peter Mell and Tim Grance, "NIST Definition of Cloud Computing v15," National Institute of Standards and Technology, October 7, 2009, pp. 1–5.

7. David Wyld, "Moving to the Cloud: An Introduction to Cloud Computing in Government," IBM Center for the Business of Government, E-Government Series, 2009, p. 12.

8. Based on discussion at Forum on Cloud Computing, Center for Strategic and International Studies, December 3, 2009.

9. Rajen Sheth, "What We Talk about When We Talk about Cloud Computing," *Google Enterprise Blog,* April 28, 2009.

10. MeriTalk, "The DIY Federal IT Bailout," February 9, 2009, p. 7.

11. Forrest, "Clearing the Air on Cloud Computing," pp. 22–24.

12. Jo Maitland, "Analysts Debate Merits of Cloud Computing for Large Data Centers," April 17, 2009 (www.searchservervirtualization.com); Steve Lohr, "When Cloud Computing Doesn't Make Sense," *New York Times,* April 15, 2009.

13. Forrest, "Clearing the Air on Cloud Computing," p. 25.

14. David Sarno, "Los Angeles Adopts Google E-Mail System for 30,000 City Employees," *Los Angeles Times,* October 27, 2009.

15. Miguel Santana, "Additional Information Requested by the Budget and Finance Committee Regarding the Contract with Computer Science Corporation for the Replacement of the City's E-Mail System," City of Los Angeles, October 26, 2009, pp. 1–2.

16. C. G. Lynch, "Washington, D.C., Uses Google Apps to Cut Government Waste," Bloomberg.com, March 10, 2009.

17. Molly Peterson, "Google Rewires Washington in Challenge to Microsoft," Bloomberg.com, October 10, 2008.

18. Kim Hart, "Google Goes to Washington, Gearing Up to Put Its Stamp on Government," *Washington Post,* September 29, 2008, p. D7.

19. C. G. Lynch, "How Vivek Kundra Fought Government Waste One Google App at a Time," CIO.com, September 22, 2008; David Sarno, "Los Angeles City Hall Becomes Tech Giants' Battlefield," *Los Angeles Times,* September, 28, 2009.

20. Microsoft Case Studies, "City Government Uses Online Services for Messaging, Saves 40 Percent Annually," Microsoft.com, May 14, 2009, p. 4–5.

21. Ibid., pp. 5–6.

22. Nucleus Research, "ROI Case Study: State Department," August 2009, p. 3.

23. David Wyld, "Moving to the Cloud: An Introduction to Cloud Computing in Government," IBM Center for the Business of Government, E-Government Series, 2009, pp. 25–26.

24. Comments and quotations in this section are from Gretchen Curtis, who was interviewed on March 15, 2010.

25. Nucleus Research, "ROI Case Study: State Department," 2009 (www.salesforce.com/assets/pdf/casestudies/State_Department_Case_Study_Final_Nucleus_Research.pdf), pp. 1–4.

26. Wyld, "Moving to the Cloud," p. 20.

27. J. Nicholas Hoover, "GSA Shifts Cloud Computing Strategy," *Information Week,* March 1, 2010.

28. Katie Lewin, "Federal Cloud Computing Initiative Overview," General Services Administration, June 18, 2009.

29. Patrick Thibodeau, "Microsoft Seeks Legal Protections for Data Stored in Cloud," *ComputerWorld,* January 20, 2010.

30. Mike Bradshaw, statement before the House Committee on Oversight and Government Reform, Subcommittee on Government Management, Organizations, and Procurement, *Cloud Computing: Benefits and Risks of Moving Federal IT into the Cloud,* July 1, 2010, p. 5.

31. Michael Goodrich, "GSA Presentation on the Federal Cloud Computing Initiative," General Services Administration, 2010, pp. 1–2.

32. Tony Romm, "GOP Hypes Skype for Congress's Computers," Politico.com, July 1, 2010, p. 13.

33. Darrell West, "State and Federal Electronic Government in the United States," Brookings, 2008, pp. 12–13.

34. Vivek Kundra, statement before the House Committee on Oversight and Government Reform, Subcommittee on Government Management, Organizations, and Procurement, *Cloud Computing: Benefits and Risks of Moving Federal IT into the Cloud,* July 1, 2010; Vivek Kundra, "State of Public Sector Cloud Computing," CIO Council, May 20, 2010.

35. "Exploring the Future of Cloud Computing," World Economic Forum report, 2010.

36. Avi Goldfarb and Catherine Tucker, "Privacy Regulation and Online Advertising," *Management Science* 57 (January 2011): 57–71.

37. Daniel Burton, statement before the House Committee on Oversight and Government Reform, Subcommittee on Government Management,

Organizations, and Procurement, *Cloud Computing: Benefits and Risks of Moving Federal IT into the Cloud,* July 1, 2010.

38. David McClure, statement before the House Committee on Oversight and Government Reform, Subcommittee on Government Management, Organizations, and Procurement, *Cloud Computing: Benefits and Risks of Moving Federal IT into the Cloud,* July 1, 2010, pp. 2–3.

CHAPTER FIVE

1. Federal Communications Commission, *Connecting America: The National Broadband Plan,* 2010, p. xi.

2. This chapter draws on Darrell West, "An International Look at High-Speed Broadband," Brookings, 2010.

3. Christine Zhen-Wei Qiang, "Broadband Infrastructure in Stimulus Packages: Relevance for Developing Countries," World Bank, 2009, p. 1.

4. FCC, *Connecting America,* p. 2.

5. Taylor Reynolds, "The Role of Communication Infrastructure Investment in Economic Recovery," Organization for Economic Cooperation and Development, Working Party on Communication Infrastructures and Services Policy, March 2009.

6. "Canada Requires ISPs to Divulge Internet Traffic Management Practices," *Government Technology,* October 23, 2009 (www.govtech.com).

7. Bobbie Johnson, "Finland Makes Broadband Access a Legal Right," *Guardian,* October 14, 2009.

8. Communications Workers of America, "Speed Matters: A Report on Internet Speeds in All 50 States," 2009 (www.SpeedMatters.org).

9. Organization for Economic Cooperation and Development, "Information Technology Outlook," Paris, 2008.

10. California Broadband Task Force, "The State of Connectivity: Building Innovation through Broadband," final report, January 2008, pp. 11–12.

11. Nielsen Company, "Internet Revolution in Full Swing with More Kiwis Accessing Broadband Than Dial-Up for First Time," press release, 2008, pp. 1–3.

12. J. C. Herz, "The Bandwidth Capital of the World," *Wired,* 2006.

13. OECD, "Information Technology Outlook."

14. Ibid.

15. "Online Computer and Video Game Industry," OECD Working Paper on the Information Economy, Paris, 2005.

16. Herz, "Bandwidth Capital of the World."

17. Christine Zhen-Wei Qiang, Carlo M. Rossotto, and Kaoru Kimura, "Economic Impacts of Broadband," in *Information and Communications for Development, 2009: Extending Reach and Increasing Impact* (Washington: World Bank, 2009), p. 41.

18. Heekyung Kim, Jae Moon, and Shinkyu Yang, "Broadband Penetration and Participatory Politics: South Korea Case," *Proceedings of the 37th Hawaii International Conference on System Sciences,* 2004, p. 4.

19. Herz, "Bandwidth Capital of the World."

20. Pew Internet and American Life Project, "Home Broadband Adoption," 2009, p. 1.

21. Reuters, "U.S. May Need as Much as $350 Billion to Extend Broadband," September 29, 2009.

22. Qiang, "Broadband Infrastructure in Stimulus Packages: Relevance for Developing Countries."

23. Strategic Networks Group, "Economic Impact Study of the South Dundas Township Fibre Network," Department of Trade and Industry, Canada, July 27, 2003, p. 3.

24. Raul L. Katz and others, "The Impact of Broadband on Jobs and the German Economy," Columbia Business School, 2009.

25. Christine Zhen-Wei Qiang, "Telecommunications and Economic Growth," World Bank, 2009, p. 1.

26. Robert Atkinson, Daniel Correa, and Julie Hedlund, "Explaining International Broadband Leadership," Information Technology and Innovation Foundation, May 2008.

27. Peter Collins, David Day, and Chris Williams, "The Economic Effects of Broadband: An Australian Perspective," paper presented at a joint WPIIS-WPIE workshop held by the OECD on May 22, 2007.

28. D. H. Shin, "Design and Development of Next Generation of Information Infrastructure: Case Studies of Broadband Public Network and Digital City," *Knowledge Technology and Policy* 18, no. 2 (2005): 101–25.

29. Katia Passerini and Dezhi Wu, "The New Dimensions of Collaboration: Mega and Intelligent Communities, ICT, and Wellbeing," *Journal of Knowledge Management* 12, no. 5 (2008): 79–90.

30. Paschal Preston and Anthony Cawley, "Broadband Development in the European Union to 2012: A Virtuous Circle Scenario," *Futures* 40, no. 9 (2008): 812–21.

31. Stefan Agamanolis, "At the Intersection of Broadband and Broadcasting: How Interactive TV Technologies Can Support Human Connectedness," *International Journal of Human Computer Interaction* 24, no. 2 (2008): 121–35.

32. Valerie D'Costa and Tim Kelly, "Broadband as a Platform for Economic, Social, and Cultural Development: Lessons from Asia," paper presented at the Joint OECD–World Bank Conference on Innovation and Sustainable Growth in a Globalized World, November 18, 2008.

33. Renate Steinmann, Alenka Krek, and Thomas Blaschke, "Analysis of Online Public Participatory GIS Applications with Respect to the Differences between the U.S. and Europe," Salzburg Research Forschungsgesellschaft, University of Salzburg, Austria, 2004.

34. Oscar Cardenas-Hernandez and Luis Martinez Rivera, "A GIS-Based Approach for Participatory Decision-Making in Mexico," Universidad de Guadalajara, Mexico, 2006.

35. G. S. Hanssen, "E-communication: Strengthening the Ties between Councillors and Citizens in Norwegian Local Government," *Scandinavian Political Studies* 31, no. 3 (2008): 333–61.

36. Tapio Hayhtio and Jarmo Rinnel, "Hard Rock Hallelujah! Empowering

Reflexive Political Action on the Internet," *Journal for Cultural Research* 11, no. 4 (2007): 337–58.

37. Hernando Rojas and Eulalia Puig-i-Abril, "Mobilizers Mobilized: Information, Expression, Mobilization, and Participation in the Digital Age," *Journal of Computer Mediated Communication* 14, no. 4 (2009): 902–27.

38. Kim, Moon, and Yang, "Broadband Penetration and Participatory Politics: South Korea Case."

39. Peter Trkman and Tomaz Turk, "A Conceptual Model for the Development of Broadband and E-Government," *Government Information Quarterly* 26, no. 2 (2009): 416–24.

40. Michael Alvarez, Thad Hall, and Alexander Trechsel, "Internet Voting in Estonia," Voting Technology Project Working Paper 60, California Institute of Technology and MIT, Pasadena, January 2008.

41. Interreg, "e-Voting Pilots (United Kingdom)," in *e-Government Actions in Europe* (Tallinn, Estonia: e-Governance Academy, 2006), pp. 51–52.

42. Darrell West, "Improving Technology Utilization in Electronic Government around the World, 2008," Brookings, 2008, p. 4.

CHAPTER SIX

1. This chapter draws on Darrell West and Jenny Lu, "Comparing Technology Innovation in the Private and Public Sectors," Brookings, 2009.

2. Ibid.

3. Ibid.

CHAPTER SEVEN

1. Darrell West, *Digital Government: Technology and Public Sector Performance* (Brookings, 2005).

2. This chapter draws on Darrell West, "What Consumers Want from Mobile Communications," Brookings, 2009.

3. Ibid., p. 2.

CHAPTER EIGHT

1. White House, "Vice President Biden Releases Report on Recovery Act Impact on Innovation," August 24, 2010.

2. This chapter draws on Darrell West, "Enabling Personalized Medicine through Health Information Technology," Brookings, 2011.

3. Jill Wechsler, "New NIH Leader Backs Personalized Medicine," *PharmExec Blog,* July 22, 2009.

4. Andrew Pollack, "Awaiting the Genome Payoff," *New York Times,* June 14, 2010; Nicholas Wade, "A Decade Later, Genetic Map Yields Few New Cures," *New York Times,* June 12, 2010.

5. Sandra Jones and Bruce Japsen, "Walgreens to Sell Genetic Tests, FDA Investigating," *Chicago Breaking Business,* May 11, 2010.

6. Claire Miller, "Bringing Comparison Shopping to the Doctor's Office," *New York Times,* June 10, 2010.

7. Clayton Christensen, Jerome Grossman, and Jason Hwang, *The Innovator's Prescription: A Disruptive Solution for Health Care* (McGraw-Hill, 2008).

8. Institute of Medicine, "The Value of Genetic and Genomic Technologies," August 2010.

9. Clifford Goodman, "Comparative Effectiveness Research and Personalized Medicine: From Contradiction to Synergy," Lewin Group Center for Comparative Effectiveness Research, October 28, 2009.

10. Ibid.

11. Gregory Downing, "Policy Perspectives on the Emerging Pathways of Personalized Medicine," *Dialogues in Clinical Neuroscience* 11, no. 4 (2009): 377–87.

12. Brian Spear, Margo Heath-Chiozzi, and Jeffrey Huff, "Clinical Application of Pharmacogenetics," *Clinical Trials in Molecular Medicine* 7, no. 5 (2001); Goodman, "Comparative Effectiveness Research and Personalized Medicine."

13. U.S. Department of Health and Human Services, "Personalized Health Care: Pioneers, Partnerships, Progress," November 2008.

14. Clay Shirky, "The Semantic Web, Syllogism, and Worldview," New York University, November 7, 2003 (www.shirky.com/writings/semantic_syllogism.html).

15. Institute of Medicine, "Beyond the HIPAA Privacy Rule: Enhancing Privacy, Improving Health through Research," February 2009.

16. Ibid.

17. Jacqueline Klosek, "Emerging Issues in Informed Consent," *Genetic Engineering and Biotechnology News,* July 8, 2010.

18. University of Pittsburgh Medical Center, "Honest Broker Certification Process Related to the De-identification of Health Information for Research and Other Duties," April 14, 2003.

19. Downing, "Policy Perspectives on the Emerging Pathways of Personalized Medicine."

20. Institute of Medicine, *A Foundation for Evidence-Driven Practice: A Rapid Learning System for Cancer Care* (Washington: National Academies Press, 2010).

21. Lynn Etheredge, "Element 3: Toward a Rapid-Learning Health System," George Washington University, April 29, 2010.

22. David Eddy, "Health Technology Assessment and Evidence-Based Medicine," *Value in Health* 12, supp. 2, 2009.

23. Jim Adams, Edgar Mounib, and Amnon Shabo, "IT-Enabled Personalized Healthcare," IBM Institute for Business Value, 2010.

CHAPTER NINE

1. This chapter draws on Darrell West, "The New Digital Press: How to Create a Brighter Future for the News Industry," Brookings, May 2009; and Darrell West, Grover J. "Russ" Whitehurst, and E. J. Dionne Jr., "Re-Imagining Education Journalism," Brookings, May 2010.

2. Walter Isaacson, "How to Save Your Newspaper," *Time,* February 5, 2009; Paul Starr, "Goodbye to the Age of Newspapers (Hello to a New Era of Corruption)," *New Republic,* March 4, 2009; John Nichols and Robert W. McChesney, *The Death and Life of American Journalism* (New York: Nation Books, 2010).

3. Pew Research Center, Project for Excellence in Journalism, "The State of the News Media, 2010," 2010, pp. 1–2 (www.stateofthemedia.org).

4. Darrell West, Grover J. "Russ" Whitehurst, and E. J. Dionne Jr., "Invisible: 1.4 Percent Coverage for Education Is Not Enough," Brookings, 2009, p. 1.

5. Darrell West, *The Rise and Fall of the Media Establishment* (Boston: Bedford–St. Martin's Press, 2001).

6. Cass Sunstein, *Republic.com* (Princeton University Press, 2001).

7. Pew Research Center, "The State of the News Media, 2010."

8. Richard Perez-Pena, "In San Francisco, Plans to Start News Web Site," *New York Times,* September 24, 2009.

9. West, Whitehurst, and Dionne, "Re-Imagining Education Journalism," p. 9.

10. All quotes in this paragraph come from Lia Miller, "New Web Site for Academics Roils Education Journalism," *New York Times,* February 14, 2005.

11. Annys Shin, "*Inside Higher Ed* Emphasizes Online Focus," *Washington Post,* March 7, 2005.

12. Joelle Tessler, "Google CEO Eric Schmidt: Newspapers Can Save Themselves with the Internet," Associated Press, April 12, 2010.

13. Ibid.

14. Frank Ahrens, "The *New York Times* Announces a Plan to Charge Readers for Online Content Starting in 2011," *Washington Post,* January 21, 2010.

15. Ibid.

16. Jessica Ramirez, "Journalism's Savior: Why Steven Brill Believes His New Company Can Save American Media," *Newsweek,* April 17, 2009.

17. Isaacson, "How to Save Your Newspaper."

18. Michael Kinsley, "You Can't Sell News by the Slice," *New York Times,* February 10, 2009.

19. Quoted in Tessler, "Google CEO Eric Schmidt."

20. Jonathan Rauch, "How to Save Newspapers—and Why," *National Journal,* June 14, 2008.

21. Nichols and McChesney, *Death and Life of American Journalism.*

22. Ibid.

CHAPTER TEN

1. Harris Interactive Survey, "Many U.S. Adults Are Satisfied with Use of Their Personal Health Information," March 26, 2007, pp. 1–2 (www.Harris Interactive.com).

2. Ivana Deyrup and Shane Matthews, "Cloud Computing and National Security Law," Harvard Law National Security Research Group, 2010.

3. This chapter draws on Darrell West, "Steps to Improve Cloud Computing in the Public Sector," Brookings, 2010; Darrell West, "Saving Money through Cloud Computing," Brookings, 2010; and Allan A. Friedman and Darrell West, "Privacy and Security in Cloud Computing," Brookings, 2010.

4. Agam Shah, "Study: Astounding Number of Laptops Lost in Airports," *Computerworld*, July 1, 2005.

5. Bruce Hoard, "8M Cell Phones Will Be Lost in '07," *Computerworld*, July 13, 2007.

6. David Molnar and Stuart Schechter, "Self Hosting vs. Cloud Hosting: Accounting for the Security Impact of Hosting in the Cloud," Microsoft Research, 2010.

7. Marjory Blumenthal, "Is Security Lost in the Clouds?" Telecommunications Policy Research Conference, October 2, 2010.

8. European Network and Information Security Agency, "Cloud Computing: Benefits, Risks, and Recommendations for Information Security," report, November 2009.

9. Blumenthal, "Is Security Lost in the Clouds?"

10. This section draws on Friedman and West, "Privacy and Security in Cloud Computing."

11. Sara Sinclair and Sean Smith, "Preventative Directions for Insider Threat Mitigation Using Access Control," in *Insider Attack and Cyber Security*, edited by Salvatore Stolfo and others (New York: Springer, 2008), chap. 11.

12. Steve Bellovin, "On the Brittleness of Software and the Infeasibility of Security Metrics," *IEEE Security and Privacy* 4, no. 4 (2006): 96.

13. Simon Bradshaw, Christopher Millard, and Ian Walden, "Contracts for Clouds: Comparison and Analysis of the Terms and Conditions of Cloud Computing Services," Queen Mary School of Law Legal Studies Research Paper 63, 2010.

14. Declan McCullagh, "Tech Coalition Pushes Rewrite of Online Privacy Law," *CNET News*, March 29, 2010.

15. Bret Taylor, statement before the Senate Committee on Commerce, Science, and Transportation, *Consumer Online Privacy*, July 27, 2010.

16. Alma Whitten, statement before the Senate Committee on Commerce, Science, and Transportation, *Consumer Online Privacy*, July 27, 2010.

17. This section draws on Friedman and West, "Privacy and Security in Cloud Computing."

18. Open Identity Exchange, "An Open Market Solution for Online Identity Assurance," March 2010.

19. Scott Charney, statement before the House Committee on Oversight and Government Reform, Subcommittee on Government Management, Organizations, and Procurement, *Cloud Computing: Benefits and Risks of Moving Federal IT into the Cloud*, July 1, 2010.

20. McCullagh, "Tech Coalition Pushes Rewrite of Online Privacy Law."

21. Miguel Helft, "Technology Coalition Seeks Stronger Privacy Laws," *New York Times*, March 30, 2010.

22. Cita Furlani, statement before the House Committee on Oversight and Government Reform, Subcommittee on Government Management,

Organizations, and Procurement, *Cloud Computing: Benefits and Risks of Moving Federal IT into the Cloud*, July 1, 2010.

23. Gregory Wilshusen, statement before the House Committee on Oversight and Government Reform, Subcommittee on Government Management, Organizations, and Procurement, *Governmentwide Guidance Needed to Assist Agencies in Implementing Cloud Computing*, July 1, 2010.

24. Helft, "Technology Coalition Seeks Stronger Privacy Laws."

25. General Services Administration, "Statement of Objectives for Enterprise E-Mail and Collaboration Services," June, 2010.

26. Bradshaw, Millard, and Walden, "Contracts for Clouds."

CHAPTER ELEVEN

1. M. C. Kintner-Meyer and others, "GridWise: The Benefits of a Transformed Energy System," Pacific Northwest National Laboratory, 2003.

2. Reuters, "U.S. May Need as Much as $350 Billion to Extend Broadband," September 29, 2009.

3. David Allen, "University Technology Transfer Effectiveness," University of Colorado, Boulder, 2010.

4. Lisa Mitchell, testimony before the House Committee on Science and Technology, Subcommittee on Research and Science Education, *Improving Technology Commercialization to Drive Future Economic Growth*, June 10, 2010.

5. Phillip Phan and Donald Siegel, "The Effectiveness of University Technology Transfer," Rensselaer Working Papers in Economics 609, April 2006.

6. Sharon Belenzon and Mark Schankerman, "University Knowledge Transfer: Private Ownership, Incentives, and Local Development Objectives," *Journal of Law and Economics* 52, no. 111 (2009): 111–44.

7. This sections draws in part on Darrell West, *Brain Gain: Rethinking U.S. Immigration Policy* (Brookings, 2010).

INDEX

ABC, 149
Adobe, 104
Advertising: online, 60, 93, 160,
161–62, 176; in traditional
media, 146, 147, 148, 164
Agamanolis, Stefan, 74
Airlines, 17
AirStrip Technologies, 31
Alford, Ted, 46, 47, 48
Alvarez, Michael, 78
Amazon.com, Elastic Compute
Cloud, 47
American Obesity Association, 38
American Recovery and Reinvest-
ment Act of 2009, 63, 128,
135
Antigua and Barbuda, Department
of Tourism, 80
Antitrust laws, 167–68
AOL, 150
Apple, 100, 113, 117
Arizona State University, 136
Associated Press, 156, 158
AT&T, 28, 71, 99–102, 176
Australia: broadband speeds, 64,
67–68; Department of Communi-
cations Information Technology,

73; immigration policy, 193;
infrastructure investments, 64
Austria, online public services, 81

Ballmer, Steven, 194
Banking. *See* Financial services
Baucus, Max, 38
Bay Area News Project, 153
Belenzon, Sharon, 192
Bienert, Phil, 99–102
Bill and Melinda Gates Foundation,
153–54, 155, 157
Blaschke, Thomas, 75–76
Blogs, 22, 23–24, 68, 98, 152, 154,
160, 162
Bloomberg News, 150
Blumenthal, Marjory, 173
Bodybugg, 28
Brill, Steven, 166
Broadband: access, 64, 65–66, 67,
71–72, 81, 119–20; applica-
tion requirements, 68, 70; civic
engagement and, 75–77, 78;
costs of improving access, 64, 72,
188–89; demand for, 68–70, 71;
economic gains, 72–73; health
care applications, 28; importance,

Comcast, 71
Community clouds, 46
Competition, 8, 104–05
Computer Fraud and Abuse Act of
1986, 179
Confidentiality. *See* Privacy
Congress, U.S.: privacy laws, 178–
79; slow acceptance of innova-
tions, 1–2, 58–59; use of social
networking sites, 22–25
Consumer Reports, 35
Cookies, 87, 91
Corporate websites: advertising, 93;
case studies, 95–105; compared
to public sector sites, 86–95, 105;
financial services, 90, 95–99; pri-
vacy policies, 91, 176–77; read-
ability levels of text, 91; services,
89–90; visitor feedback, 94. *See
also* Websites
Corventis, 28
Courts. *See* Judiciary
Cross-border computing, 57, 60–61
Crovitz, Gordon, 166
Crowd sourcing, 34
Current Population Survey, 119–20
Current TV, 151
Curtis, Gretchen, 53, 54–55
Customers: feedback from, 7–8, 9,
94; focus on, 7–8; of government
agencies, 100–01, 104; satisfaction, 8–9, 21–22

Data.gov, 19
Data storage, 32, 47, 51–52, 58, 174.
See also Electronic medical records
Day, David, 73
D'Costa, Valerie, 75
Defense, Department of, 20, 21
Demcom.house.gov, 23
Democracy: citizen communication
with legislators, 24; informed citi-
zenry, 169; online voting, 78–79;
public participation, 19–20; role
of media, 147

Digital cities, 74
Dill, Clarence, 1
Disability access to websites, 87, 90,
91–92
DNA. *See* Genetic sequencing
Duke University, 143
Dynamed Solutions, 30

Economic growth: broadband penetra-
tion and, 3–4; role of innovation,
3–4; stimulus packages, 64, 66
ECPA. *See* Electronic Communica-
tions Privacy Act
Ecuador, Ministry of Defense, 80
EdNews.org, 151
Education: blogs, 152; job advertis-
ing, 161–62; media coverage,
146, 148, 153, 154–63; science,
engineering, math, and technology, 189; websites focusing on,
151, 152, 155–57, 160–63
Education Week, 153, 155–57
Edwards, Virginia, 155–57
Efficiency, 16–18, 31
E-government, 78
Egyptian Tourism website, 80
Elections, 78–79
Electronic Communications Privacy
Act (ECPA), 178
Electronic medical records: benefits,
40, 131–32; data coding, 132–34;
genetic sequencing data, 128, 134;
obstacles to data sharing, 132–33,
173; online storage, 32, 127; privacy issues, 134–39; public fund-
ing, 128; public support, 31
E-mail: cloud computing, 46, 49–50,
53; health care reminders, 29–30,
31, 36, 127; local government
use, 76; security issues, 180
Energy, smart grids, 4, 187–88
Estonia, online public services, 78
European Union: broadband policy,
74; privacy and security issues, 60
Evans, Helen Hughes, 40

Facebook, 22, 23–24, 145, 176
Farbert, Amy, 34
Federal Communications Commission (FCC), 63, 64, 71–72, 119, 188–89
Federal government: cloud computing, 52–56, 62; contracting and certification processes, 59–60; courts, 59, 180–81; data centers, 45, 48, 58; immigration policy, 192–94; information technology policy, 57–62; information technology spending, 45; online data, 19; organizational changes, 17–18; procurement, 56, 59–60; websites, 91. *See also* Congress; Government agencies
Federal Information Security Management Act, 56
Federal Risk and Authorization Management Program, 60
Federal Trade Commission (FTC), 172
FedEx, 102–05
Fiji, online public services, 80
Financial services: technological innovation, 16; websites, 90, 95–99
Financial Times, 166
Finland: broadband access, 72; broadband speeds, 64, 67–68; infrastructure investments, 64; online civic engagement, 76–77
Fleming, Russ, 102–05
Florida, privacy laws, 138
Forman, Mark, 46
Forrest, William, 47, 48
Foundations, 153–54, 155, 156, 157–58
Fourth Amendment, 180
France: broadband access, 67; health care spending, 39
FTC. *See* Federal Trade Commission

Gates Foundation, 153–54, 155, 157
Geisinger Health System, 143
Geisinger Medical Center, 36–37

General Services Administration (GSA), 21, 56, 62, 179–80
Genetic conditions, 133
Genetic Information Nondiscrimination Act, 135
Genetic sequencing: applications, 129–30; costs of testing, 139; human genome map, 125, 128–30; interpretation of tests, 140; privacy issues, 135, 136, 138–39
Geographic information systems (GIS), 75–76
Germany: broadband speeds, 64, 67, 68; economic gains of investments in broadband, 73; interactive mapping, 76
GIS. *See* Geographic information systems
Glass, Carter, 1
Gluco Phones, 28–29
Goldfarb, Avi, 60
Google, 32, 49, 113, 117, 150, 173
Government: dysfunctional politics, 15–16; efficiency and productivity, 16–18; information technology spending, 45; lack of trust in, 15, 195; reform approaches, 17–18. *See also* Federal government; Local governments; State governments
Government Accountability Office, 179
Government agencies: cloud computing, 48, 179–80; creating, 17–18; customers, 100–01, 104; online services, 18–19, 77–81; organizational structures, 17, 185–86; performance expectations, 194–95; public participation in rule making, 19–20; reorganizations, 17; responsiveness, 20–22; websites, 21–22, 32, 78, 80–81, 89, 90–91. *See also* Public sector innovation
Greenberg, Steven, 29
Green, David, 30

INDEX

Grimshaw, Rob, 165
Groark, Marie, 155
GSA. *See* General Services
Administration

Hall, Thad, 78
Hanssen, G. S., 76
Harkin, Tom, 38
Havasupai tribe, 136
Hayhtio, Tapio, 76–77
Health and Human Services, Department of (HHS), 17, 32, 135, 142, 144
Health care: broadband speeds needed, 68; changes in system, 27–28, 35–38; clinical use of research results, 140–41; coordination, 36–37; crowd sourcing, 34; current system, 26, 126; defensive medicine, 40; delivery, 126–27; efficiency, 31; incentives, 35–38, 142; news, 153, 154; online information, 32–33, 34, 127, 151; outcomes, 37, 39; predictive modeling, 141–43; quality, 26, 34–35; revolutions, 126–30; spending on, 39. *See also* Personalized medicine
HealtheTrax software, 30
Health information technology: benefits, 143; challenges, 130–43; cost savings, 39–40; criticism of, 40; data coding, 132–34, 140, 141; data integration, 125–26, 130–34, 141; electronic medical records, 31, 32, 40, 128, 131–39, 173; innovations, 26–35, 127–28; mobile health, 26–27, 127; for personalized medicine, 125–26; remote monitoring devices, 28–29, 36, 39–40, 127; use of, 39; websites, 32–33, 34, 127, 151
Health insurance: premiums, 38; reimbursements, 27, 36–37, 126, 139–40; use of genetic information, 135

Health Insurance Portability and Accountability Act (HIPAA), 134–35, 138
Hellman, Warren, 153
Henneberger, Melinda, 150
Heyward, Andrew, 45
HHS. *See* Health and Human Services, Department of
Higher education, 157, 160–63, 189, 190–92. *See also* Education
High-speed broadband. *See* Broadband
Hindery, Leo, Jr., 166
HIPAA. *See* Health Insurance Portability and Accountability Act
Hospitals, 32, 34–35, 127
Huffington Post, 154
Human genome map, 123, 128–30
Hybrid clouds, 46

IaaS. *See* Infrastructure-as-a-service
Imager Software, 51–52
Immigration policy, 192–94
Incentives: in health care, 35–38, 142; promoting innovation, 8, 186–87, 192
Indiana, state government websites, 19, 90
Information technology: federal policy, 57–62; spending on, 7, 45. *See also* Cloud computing; Health information technology; Innovation
Infrastructure: Internet as, 3; investments in, 64. *See also* Broadband
Infrastructure-as-a-service (IaaS), 53, 56
Innovation: adoption, 187–88; advantages, 2, 3–4; commercializing university research, 190–92; disruptions, 1–2; economic gains, 3–4; facilitating, 185–95; factors in successful, 6–9; historical, 5; incentives, 8, 186–87, 192; risks, 4–5. *See also* Broadband; Cloud

213

Starr, Paul, 145
State, Department of, Nonprolifera-
tion and Disarmament Fund, 52
State governments: information tech-
nology budgets, 7; privacy laws,
138–39; websites, 19, 90–91. *See
also* Public sector innovation
Steinmann, Renate, 75–76
Stimulus packages, 64, 66
Strategic Networks Group, 72
Sunstein, Cass, 151
Supreme Court, 59
Sweden, broadband speeds, 67
Systematized Nomenclature of Medi-
cine Clinical Terms, 133

Taiwan, online public services,
77–78
Tax credits: for employer health
programs, 38; for news subscrip-
tions, 168–69; for research and
development, 189
Technological innovation. *See*
Innovation
Telecommunications markets: competition, 110; regulation, 109. *See
also* Broadband
Telefonica, 109
Telephones, rotary dial, 1–2. *See also*
Cell phones
Television news, 5, 149. *See also*
Traditional media
Text messaging, 30
Time magazine, 145
T-Mobile, 71
Traditional media: advertising revenue, 146, 147, 148, 164; changes,
145–46, 147–49; education coverage, 146, 148, 156; financial
pressures, 145, 146–47, 149–50;
political coverage, 23; public
broadcasting, 147, 153–54, 155,
157–58, 159; radio, 5; television
news, 5, 149; wire services, 156,
158. *See also* Newspapers

Transparency: of cloud computing
performance, 61–62; of govern-
ment, 23; of privacy practices,
176–77
Trechsel, Alexander, 78
Triage Wireless, 28
Trkman, Peter, 78
Tucker, Catherine, 60
Turk, Tomaz, 78
Twitter, 22, 23–24, 98, 145
Tydings, Millard, 1

United Kingdom: health care spending, 39; immigration policy, 193;
interactive mapping, 75–76; local
governments, 78–79; mobile
technology market, 109, 110–13;
online voting, 78–79
United States: consumer behavior
and attitudes survey, 113–18;
health care spending, 39; infrastructure investments, 64; mobile
technology market, 109, 110–13,
114, 115. *See also* Federal
government
U.S. Air Force, cloud computing, 55
University of California at Santa
Barbara, 53
University of Pittsburgh Medical
Center, 136–37
USA.gov, 62
USA Today, 162
U.S. Steel, 89–90
Utah State University, Center for
Persons with Disabilities, 92

Venture capital, 191, 192
Verizon, 71, 90
Virtual reality, 70, 74–75
Von Finckenstein, Konrad, 66–67
Voting, online, 78–79

Wall Street Journal, 165
Washington, D.C., cloud computing,
50–51